THE HISTORY OF HEAVEN

About the Cover Image

A computer simulation has reproduced an astronomical event recorded in the *Dangunsegi* as having occurred in mid-July of 1734 BCE : the alignment of Mercury, Venus, Mars, Jupiter, and Saturn in the western sky. The existence of this verified astronomical record in itself proves that Dangun Joseon was an actual country with an astronomical observatory—not a myth, but a country in historical fact.

Copyright © 2022 by Dragon and Phoenix Publishing

All rights reserved.
Published in the United States by Dragon and Phoenix Publishing.
No part of this publication may be reproduced, distributed, or transmitted in any form or by any means, including photocopying, recording, or other electronic or mechanical methods, without the prior written permission of the publisher, except in the case of brief quotations embodied in critical reviews and certain other noncommercial uses permitted by copyright law.

For permission requests, write to the publisher, addressed "Attention: Permissions Coordinator," at the contact information below.

Dragon & Phoenix Publishing
132 Brookside Avenue
Cresskill, NJ 07626

Publisher's Cataloging-in-Publication data

Written by Seok Jae Park, Ph. D.

The History of Heaven

ISBN: 978-1-7378756-3-5
1. Korean History, 2. Korean Cosmology, 3. Korean Culture, 4. Korean Philosophy

First Edition

Sources of images:
Korea Astronomy and Space Science Institute Sangsaeng Television Broadcasting Daehan History and Culture Association

A Brief Introduction to Essential Concepts in the East and West

THE HISTORY OF HEAVEN

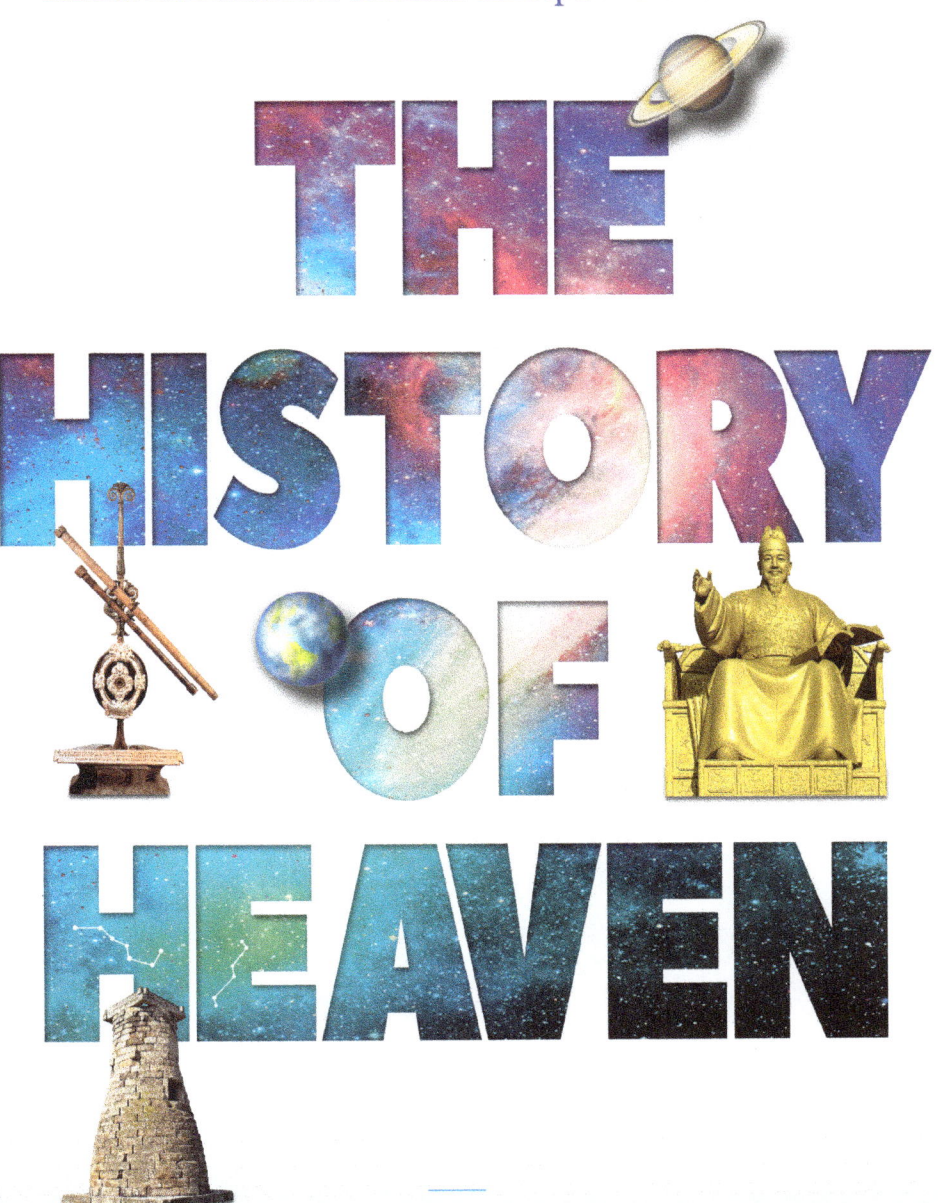

Seok Jae Park

Dragon & Phoenix Publishing

Contents

Prologue _06

Chapter 1. The Sun and Moon in the Beginning

    The sun and moon _10
    Bokhui's cosmos _15
    Aristotle's cosmos _18
    Lunar calendar and solar calendar _22

Chapter 2. Ancient Secrets carried by the Zodiac

    Geocentric view of the universe _28
    28 Constellations of the East _34
    12 zodiac constellations of the West _38
    Big Dipper and Southern Cross _41

Chapter 3. Oriental Cosmos Worked with Taegeuk

    Cheonbugyeong _48
    Four signs and eight trigrams _53
    Bokhui's eight trigrams _58
    Hado _63
    Nakseo and King Wen's eight trigrams _69

Chapter 4. Our Heaven Opened with Gaecheon

    Dangun-Joseon as history, not myth _78
    Lineage of Korean Sovereignty in BCE _82
    Korean history began with Gaecheon _87
    Emergence of Juyeok _93

Chapter 5. Cheonsang yeolcha bunya jido, Korea's Crowning Achievement

Astronomy during the Three Kingdoms period and Goryeo dynasty _102
Cheonsang yeolcha bunya jido _107
King Sejong the Great _112

Chapter 6. Astronomical Renaissance of Modern Europe

Heliocentric view of the universe _120
Discovery of gravity _124
Completion of lunisolar calendar _129

Chapter 7. The Great Universe Unveiled

Uranus, Neptune and Pluto _136
From Juyeok to Jeongyeok _140
Great universe unmasked as late as 100 years ago _145

Chapter 8. The Theory of Relativity Changing Our Concepts of Time and Space

Space contraction and time dilation _156
Physics of curved time and space _162
Black hole, the ugly duckling _166
Universe born with the Big Bang _170

Chapter 9. Quantum Physics Bringing Light to Atomic Energy

Nuclear fusion, the powerhouse of stars _178
Black hole becomes swan _184
Matter and vacuum _188
The beginning and the end _193

Epilogue _201

Prologue

In 2013, I gave a series of lectures titled "The History of Heaven" as a part of EBS (Educational Broadcasting System)'s special history lecture program. You can watch them on EBS's website or YouTube. This book shares its content and titles with those 10 lectures, but much has been added. I wrote this book based on the following two principles:

First, the East and West should be approached with equal weight. Until now, the history of the universe has been described from the Western perspective. As a result, too many people still believe that Eastern traditions are unscientific and wrong while Western ones are scientific and right. This prejudice produces blind Western-centrism and adversely affects every field.

One of the most salient examples can be seen in schools where students learn Aristotle's four elements in the Western tradition but not Bokhui's five elements in the Eastern one. As a result, a majority of Koreans did not learn the principle of Taegeukgi, the national flag of Korea. It is a lamentable fact that most of us do not exactly understand our national flag.

Unlike Aristotle's four elements, Bokhui's five elements not only physically constitute the universe but also spur its continued evolution through their chemical mutual life-giving and mutual conflict. Due to this fundamental difference, the cosmic view of Bokhui greatly impacted the spiritual culture of the East. We can say that the Aristotelian conception of the universe remains a sort of physical depiction, while that of Bokhui reaches the metaphysical realm.

Second, the universe and cosmos should be approached harmoniously. The world we live in is expressed as either universe or cosmos. The term "universe" refers to the vast world filled with stars and galaxies. For example, if a book is titled *Universe*, it is very likely to be an astronomy textbook. It is difficult to associate the term "universe" with non-scientific subjects like the humanities.

The cosmos is "the universe plus something else." Here, "something else" refers to humans' subjective demands. For example, Badook (Go in Japanese) players say that the Badook board is the cosmos. Religions also say various things about the cosmos. Until the medieval period, people had talked about the cosmos rather than the universe.

In some cases, the cosmos is more important to our lives than the universe. This is especially so among Asians. The Asian game Janggi can be compared to chess, played in the West. However, Badook has no counterpart in the West. This makes the difference between East and West remarkable. This book is titled *The History of Heaven* rather than *The History of the Universe* because it deals with both the universe and the cosmos.

To help you enjoy this book, I will be employing the cosmic god and his disciples as assistants. I published the cartoon series *Cosmic God and His Disciples* in a Kids' Science Journal.

The cosmic god is the supreme god and takes charge of the structure and evolution of the universe. He wears a Taegeuk symbol upon his chest. The galaxy god manages the realm of stars and galaxies. He looks like a nerd but is the most distinguished disciple of the cosmic god. He wears a galaxy symbol upon his chest. The earth god manages the earth and the solar system. As the youngest, he takes on many unpleasant tasks. Together with these three jolly gods, let's study the history of heaven!

Seok Jae Park

Chapter 1

The Sun and Moon in the Beginning

In the past, people believed that the sun revolved around the earth once a year. The word for sun is "hae" in Korean. Accordingly, one year and two years are expressed as one "hae" and two "haes." Thirty days — that is, the time required for the moon to revolve around the earth — is called one "dal." "Dal" means month and also signifies the moon. In English, "moon" and "month" share the same root, though there is no such relation between "sun" and "year."

The sun and moon

In the Eastern tradition, the moon represents "yin (음)" energy and the sun "yang (양)" energy. Accordingly, the moon and the sun are called "tae yin (태음)" and "tae yang (태양)," respectively.

One of the most amazing facts is that the sun and the moon appear the same size. There is perhaps no similar case in the universe. This is because the sun is 400 times larger than the moon but also 400 times farther away from the earth. What a coincidence! Just imagine if the sun looked twice the size of the moon from the earth. If that were the case, the sun and the moon would not be treated equally.

As the sun and the moon appear the same size, yin and yang in the Eastern tradition are considered on an equal basis. That is, yin and yang do not conflict with each other, with one representing good and the other bad; rather, they are complementary. The sun and the moon have an enormous effect on humans. If there were two suns over our heads, things might have been astonishingly different. If the sun rose in the west and set in the east, the course of world history might have been very different.

Meanwhile, the Western tradition adopted the idea that days and nights are dominated by gods and devils, respectively. For this reason, the moon, representing the night, came to have a negative image. In Latin, the sun is "Sol" and the moon is "Luna." The word lunacy refers to mental illness, and the root of this word is luna. In the same vein, we can easily grasp how the word "moonstruck" refers to a kind of madness.

The difference between the East and the West in this matter, which goes back to time immemorial, is still influential today. People in the

East welcome a full moon, which is considered to be a time of suicide by Westerners. A full moon in the Western tradition evokes a sense of horror. For example, when a full moon rises on Friday the 13th, people tend to refrain from going out. In most Western ghost stories, ghosts appear or men turn into werewolves under the full moon. On the other hand, the full moon has a positive image in the Eastern tradition. For this reason, Eastern ghosts come out only on nights of the dark moon.

The moon may be the most friendly of all the heavenly bodies. The moon touches our sentiments better than the shining sun of the day. The moon often appears in Eastern traditional paintings, but the sun hardly does. The surface of a full moon always looks the same, as shown in Figure 1-1. This is because the moon rotates just once as it completes a revolution around the earth.

Figure 1-1 Full moon in the early evening

Since childhood, we in Korea have known of the story that a rabbit is milling on the moon. If you draw a line around the black area, it looks like a rabbit milling with a mortar and pestle, as shown in Figure 1-2.

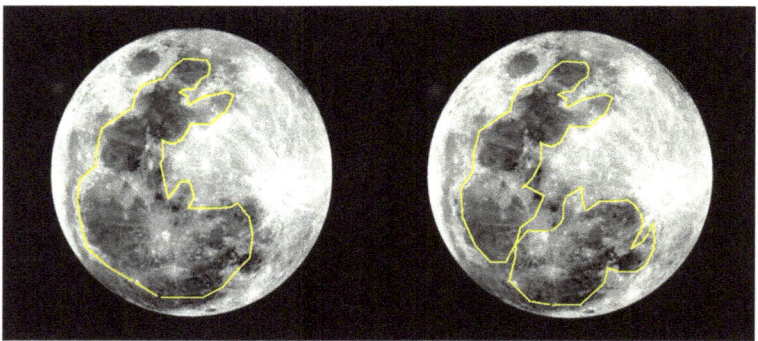

Figure 1-2 The milling rabbit (left) and the rabbit with the toad (right)

Both Figure 1-1 and Figure 1-2 show a full moon rising over a hill in the early evening. As the moon rotates clockwise over time, when it reaches the highest point in the sky around midnight, the rabbit image is upside down, as illustrated in Figure 1-3. Then the moon moves to the right. In the early morning, the full moon sets over a western hill, and the rabbit is located at the bottom.

Figure 1-3 Full moon at midnight (left) and in the early morning (right)

The full moon rotates throughout its cycle but cannot turn upside down. We see many works of art that look awkward due to a mirrored full moon. A similar case occurs with the Big Dipper (북두칠성). For example, Figure 1-4 is a mirror image of Figure 1-1, with the left and right reversed. We cannot obtain Figure 1-4 by rotating Figure 1-1. In other words, there is no full moon like Figure 1-4. There are also many examples of works that reference an image of the "dark side" of the moon as photographed by a spacecraft.

Figure 1-4 Moon that is horizontally reversed

If we observe the moon changing from a waxing crescent moon to a waning crescent moon over a month, we see that its appearance changes in the pattern depicted in Figure 1-5. The seven moons in Figure 1-5 appear around the 3rd, 8th, 11th, 15th, 19th, 22nd and 27th days of the lunar month, respectively.

The first-quarter half moon shows the rabbit and the third-quarter half moon shows the mortar. Interestingly, the moon changes shape in a different order in the southern hemisphere, that is, below the equator.

For example, in Australia, the moon's shape changes in the order from right to left in Figure 1-5.

Figure 1-5 Shapes of changing moon

A solar eclipse occurs when the moon moves between the earth and the sun. The solar eclipse occurs only on the last day of the lunar month, where the moon does not appear at night. A total solar eclipse occurs when the moon completely covers the sun. A partial solar eclipse occurs when the sun is partially covered by the moon. A total solar eclipse occurs for only a few minutes, while a partial solar eclipse can last several hours. We can experience eclipses because the sun and the moon look the same in size. The corona, which is visible only during a complete solar eclipse, is named for the Latin word for crown. This term is used because a total solar eclipse produces a bright crown shape.

A lunar eclipse occurs when the moon moves into the earth's shadow. Accordingly, the lunar eclipse appears only when the moon is full. As the earth's shadow is larger than the moon's, both total and partial lunar eclipses last for several hours. It was only a few hundred years ago that people came to accurately foretell lunar and solar eclipses.

Bokhui's cosmos

The sun and the moon may have inspired Eastern people to devise the cosmic view of yin and yang. According to this view, everything in the universe has a counterpart. If males are yang, then females are yin. If the sky is yang, then the earth is yin. If the day is yang, the night is yin. In this manner, the entire universe is composed. Such a view is characteristic of the East and not found in the Western tradition.

The cosmic view of yin and yang can be expanded so that other pairs like brightness/darkness, up/down and front/back are understood. Furthermore, yin and yang change in mutual interactions. A male and a female produce children, and the heaven and the earth bring forth everything. Yang becomes stronger as the day breaks in the early morning and reaches its peak, and then yin grows until midnight.

Thus, yin and yang are energies creating and evolving the universe. In other words, the harmony between yin and yang is a necessary and sufficient condition to maintain the universe. Moreover, the cosmic view of yin and yang goes beyond the physical realm and has an effect on the spiritual realm of human beings. As each person has yin and yang energies inside their body, their well-being requires the harmony of yin and yang with everything outside the body.

Taeho Bokhui, the founder of the cosmic view of yin, yang and the five elements, made the first Taegeukgi (Korean national flag, 태극기) about 5,500 years ago during the era of Hwanung Baedal. Thus, the Korean national flag traces back 5,500 years. Do you know any other country whose national flag goes back over 5,500 years? I am sure no country, even among the four major cradles of human civilization, has such a historic flag. Moreover, Taegeukgi is the only flag that is designed based on the principles of the universe.

It is said that Bokhui was so well-versed in astronomy as to distinguish 24 seasonal divisions.

Spring	Ipchun, Woosoo, Gyeongchip, Chunboon, Cheongmyeong, Gokwoo
Summer	Ipha, Soman, Mangjong, Haji, Soseo, Daeseo
Autumn	Ipchoo, Cheoseo, Baekno, Chuboon, Hanno, Sanggang
Winter	Ipdong, Soseol, Daeseol, Dongji, Sohan, Daehan

Table 1-1 24 seasonal divisions.

Bokhui's cosmology of yin, yang and the five elements is a much more elaborate and meaningful system that that of Aristotle's four elements. The four elements of Aristotle are mere physical constituents of the world, while Bokhui's five elements — wood (목), fire (화), earth (토), metal (금), and water (수) — have mutual life-giving (상생) and conflict (상극) relationships that also make the universe evolve chemically, as shown in Figure 1-6.

Bokhui

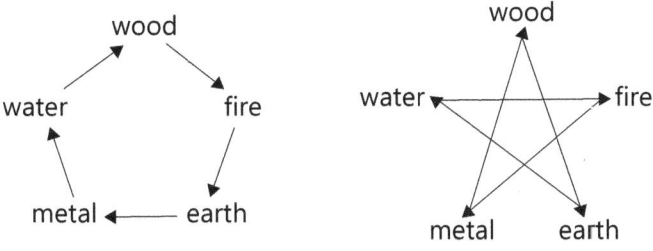

Figure 1-6 Bokhui's five elements

The five elements of Bokhui circulate in the order of wood → fire → earth → metal → water. The first change from wood to fire shows that fire is given life by burning wood. Likewise, the next change from fire to earth means that earth is given life by fire. The change from earth to metal and from metal to water show that metal comes from earth and water from metal, respectively. The last pair of elements, water and wood, indicates that water gives life to wood. The circulation of mutual life-giving is realized in this order: water → wood, wood → fire, fire → earth, earth → metal, and metal → water. Here, the arrow means to give life. Mutual conflict is obtained by skipping the very next element in the process of mutual life-giving. Thus, the circulation of mutual conflict is described in the following order: water → fire, fire → metal, metal → wood, wood → earth, and earth → water. Here, the arrow means to overcome.

Aristotle's cosmos

The early astronomical observations of primitive people laid a foundation for the scientific achievements of the ancient Greek natural philosophers. This period, from the 6th to the 7th century BCE, did not separate science from religion and philosophy. The primary interest of the natural philosophers was "arche," that is, the fundamental substance of the universe. Thales argued that water is the primary substance of nature, while Heraclitus and Anaximenes selected fire and air, respectively.

The most remarkable of the natural philosophers was Pythagoras, who is renowned for the Pythagorean theorem. He was almost obsessed with the belief that the universe is made of beautiful arrangements with perfect order. He thought that every beautiful or harmonious thing was composed of a numerical order. The achievements of Pythagoras that are known to us are so numerous that we still wonder whether they were made by him alone or the Pythagorean school and attributed to the legendary figure.

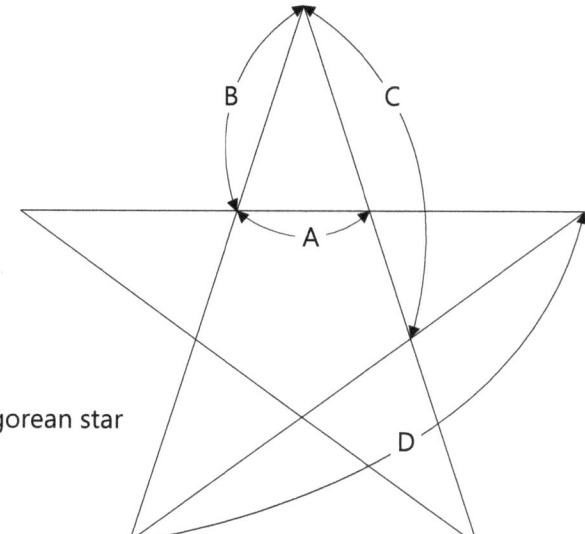

Figure 1-7 Pythagorean star

The star in Figure 1-7 is known to be made by Pythagoras. We commonly call this figure star. The Pythagorean school was so fond of this symbol that they inscribed it on the palms of their hands to use as a pass. Why did they love this star? Because the four lengths of A, B, C and D in Figure 1-7 are in the golden ratio to one another, as follows.

$$A : B = B : C = C : D = 1 : 1.618$$

To the members of the Pythagorean school, the star was the most beautiful figure in the world. Perhaps this is why stars are so often used for military insignias, for example, to rank generals (one-star, two-star, three-star, etc.). As illustrated in Figure 1-8, the golden section refers to a ratio at which an infinite number of squares can be drawn in sequence. In other words, the sides A, B, C, D, etc. of squares satisfy the following condition

$$A : B = B : C = C : D = \ldots\ldots = 1.618 : 1$$

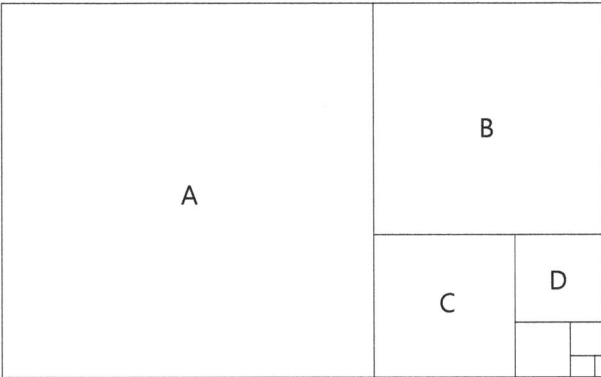

Figure 1-8 Definition of the golden section

Aristotle synthesized the cosmological arguments of the natural philosophers and proposed the theory of four elements. He argued that everything in the universe is a mixture of water, fire, air and earth in a certain ratio.

Aristotle

As earth sinks in water and air bubbles float, fire burning in air is lightest among the four elements, followed by air, earth and water. Consequently, in the Aristotelian universe, no motion can exist when earth, water, air and fire are arranged in order from below.

However, the universe is actually in flux, since the elements are mixed chaotically. For example, steel, which primarily comprises earth, inevitably falls in the air. According to the Aristotelian theory of motion, the weight is a sort of measure for the tendency to fall to the center of the earth; thus, the heavier the object, the faster it should fall. This mistaken idea had been accepted and taught without question until Galilei dropped two objects of the same size but different masses from the top of the Leaning Tower of Pisa. In this famous experiment, Galilei demonstrated that the two objects fall on the ground at the same time.

Lunar calendar and solar calendar

It takes one year for the earth to revolve around the sun. It takes one month for the moon to revolve around the earth. Naturally, one year corresponds to about 360 days, and one month to about 30 days. For this reason, although the ordinary person has 10 fingers and 10 toes, one year consists of 12 months. The 360 degrees of a circle originated from the revolution of the earth. Thus, the numbers 12, 30 and 360 are the codes of the universe revealed, inherited from primitive times.

As there are 29.5 days between two full moons, one year in the lunar calendar is 354 days (= 29.5 days × 12). There is a difference of 11 days between one year in each calendar (365 − 354 = 11). For three days, this gap becomes almost a month after three years. A leap month occurs almost every three years to compensate for the discrepancy between the lunar calendar and the solar calendar. However, things are not as simple as they seem.

Meton, an astronomer in the 5th century BCE, found out that seven leap months in every 19 years make the lunar calendar fit into the solar calendar. The solar calendar we use today was established in Rome, but its history goes back to before Meton.

Romulus, who founded Rome in the 8th century BCE, made a calendar starting from spring. One year comprised only 10 months and 304 days. As shown in Table 1-2, December, the last month, was the 10th month. The word December contains "Deca," meaning ten. Thus, it was natural enough that December was the 10th month. As one year in the Romulus calendar was very short, the introduction of leap months was inevitable. Of course, the leap month of the solar calendar is totally different from that of the modern lunar calendar. The current solar calendar has only a leap year, but not a leap month.

Numa, Romulus's successor, added two months, which are the first two months shown in Table 1-2. Under this system, one year became 355 days. Interestingly, the lengths of the months ranged irregularly between 29 and 31 days, and only February had 28 days. However, leap months were added more irregularly or for political purposes, which caused many problems.

Romulus			Numa			Julius		
month	name	days	month	name	days	month	name	days
1	Martius	31	1	Janualis	29	1	Janualis	31
2	Aprilis	30	2	Februalis	28	2	Februalis	29
3	Maius	31	3	Martius	31	3	Martius	31
4	Junius	30	4	Aprilis	29	4	Aprilis	30
5	Quintilis	31	5	Maius	31	5	Maius	31
6	Sextilis	30	6	Junius	29	6	Junius	30
7	September	30	7	Quintilis	31	7	Julius	31
8	October	31	8	Sextilis	29	8	Sextilis	30
9	November	30	9	September	29	9	September	31
10	December	30	10	October	31	10	October	30
11	—	—	11	November	29	11	November	31
12	—	—	12	December	29	12	December	30
		304			355			365

Table 1-2 Roman solar calendar

When Julius, who is known to us by the famous name of Caesar, took over, he removed leap months and employed a new calendar in which one year was 365 days. In astronomy, the exact duration of one solar year is 365 days, 5 hours, 48 minutes, and 46 seconds, or 365.2422 days. Almost one day is added every fourth year ($0.2422 \times 4 = 0.9688$ days).

To compensate, the Julian calendar introduced a leap year with 366 days every four years, and one day thus added was February 29th. As shown in Table 1-2, each month in the Julian calendar alternates between 30 and 31 days. The odd-numbered months were set to each last 31 days. Julius changed the seventh month, Quintillis, to a version of his own name. In English, it is known as July. He ordered that this calendar be adopted from the New Year in 46 BCE.

Cosmic God and His Disciples: Pleasant Earth

Galaxy god : "As your planet Earth is the most pleasant place in the universe, I invited our master here."
Earth god : 'I hope he likes it…'

Cosmic god: "By the way, the moon rabbit does not look happy."
Earth god : 'Poor rabbit, his arm must hurt after milling
 for so long…'

Chapter 2

Ancient Secrets carried by the Zodiac

From time immemorial, people could distinguish stars and planets. In addition, when it was learned that the earth is round, the geoocentric view of the universe emerged. In the Western tradition, the 12 signs of the zodiac were established, which are the constellations that hide behind the sun during the day. In the Eastern tradition, 28 constellations were determined based on the blue dragon, the white tiger, the red phoenix and the black turtle. At night, the Big Dipper and the Southern Cross became a sort of clock in the northern and southern hemispheres, respectively.

Geocentric view of the universe

Objects in the sky are generally classified into planets and stars. In English, this distinction is clear enough. The planets orbit around the sun. The solar system currently includes eight planets — Mercury, Venus, Earth, Mars, Jupiter, Saturn, Uranus and Neptune — and the dwarf planet Pluto. Unlike stars, planets do not give off their own light, but reflect sunlight. Nevertheless, as the planets are much closer to the earth than countless other stars, they look like bright stars.

The Korean term for planet is "haengseong"(행성); "haeng" means "to wander." What does it mean that the star wanders? Imagine a constellation including one bright star. If you find that the star has moved after one month but the other ones have stayed in the same place, then that bright star is a planet.

Mercury, Venus, Mars, Jupiter and Saturn are especially bright and thus are called the "Five Planets (오성)" in the East. On the other hand, Uranus and Neptune are not bright enough to be seen with the naked eye. In the Eastern tradition, the planets were named wood, fire, earth, metal and water based on the principles of yin, yang and the five elements. The ancients must have marveled at the coincidence between the five elements and the number of planets visible in the sky. Only Mars seems to match its element, fire, since it looks red, but the names of the other planets convey no particular feature.

From the astronomical perspective, the principles of yin, yang and the five elements were ideally realized by the setting of the sun, the moon and the five planets. The days of the week are also named after Sunday (일) symbolizing the sun, Monday (월) evoking the moon and the five planets Tuesday (화), Wednesday (수), Thursday (목), Friday (금) and Saturday (토).

As the five planets were visible to the naked eye, they were studied independently in the East and the West. Accordingly, the names Mercury, Venus, Mars, Jupiter and Saturn have no relation to the names used in the East. All the English names originate from Greek gods.

Mercury has such a short orbital period that it appears in the morning for two months and then in the evening for two months. That is why this planet was named after Mercury, the messenger of gods. Venus is so bright and beautiful that its name coincides with the goddess of beauty. Mars, the god of war, became the name of the planet that looks red. Jupiter, the king of gods, and Saturn, the father of Jupiter, were assigned to the remaining two planets.

English	Korean	Greek	Sign
Mercury	수성	Hermes	☿
Venus	금성	Aphrodite	♀
Mars	화성	Ares	♂
Jupiter	목성	Zeus	♃
Saturn	토성	Chronos	♄

Table 2-1 Five planets

Primitive people could distinguish planets from stars through constant observation. However, they could not determine that the earth is round. Accordingly, both in the East and West, people might have imagined a common view of the universe — that the earth is flat and covered by a semi-circular sky overhead. As they could see the difference between the planets and stars, I suppose the Easterners described the universe as in Figure 2-1. The names of the planets are given in old Korean characters.

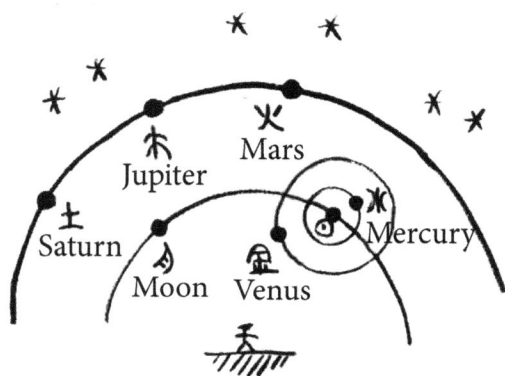

Figure 2-1 Ancient eastern view of the universe: schematic image
(Source: Gaecheongi)

Ancient Westerners such as Thales already knew that the earth is round. It is reasonable to guess that the sages in the East around the same time also realized this truth, since human intelligence is similar across humanity. Unfortunately, as there is no record in the East, we have nothing to add here regarding the Eastern tradition.

In any case, after realizing that the earth was round, one reasonable cosmological model was proposed. The Aristotelian universe includes only vertical linear motions. However, in addition to the moon orbiting around the earth, all the planets also appeared to circle around it. For this reason, Aristotelian cosmology had to suppose different dominating principles above and below the moon. The existing four elements seemed to be unsuitable for making up the planets above the moon. Thus, Aristotle added the fifth element which he called "ether."

When Alexander the Great conquered a vast portion of the world, Alexandria in Egypt naturally became the hub of the academic world. The Aristotelian cosmology and the ancient knowledge of astronomy flowed into the city, where geocentric cosmology was born at last. Ptolemy in his *Almagest* argued for a geocentric universe as shown in Figure 2-2, in which the earth is at the center of the universe.

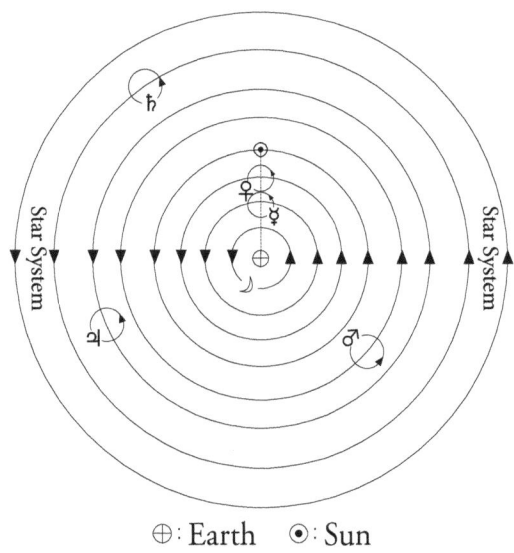

⊕: Earth   ⊙: Sun

Figure 2-2 Geocentric universe

In the geocentric universe, the planets are attached to celestial crystal spheres made of ether. As the celestial spheres revolve at different speeds, the planets travel across the night sky. In other words, the heaven consists of multiple layers. Similarly, the Eastern tradition has the idea of nine heavens (구천).

If you look carefully at Figure 2-2, you may find that the planets are moving in smaller circles. In this way, the retrograde motions of planets, which are temporary westward motions opposite to those of other planets, could be explained. The Aristotelian theory of motion and the geocentric cosmology of Ptolemy dominated the Western world for 1,400 years, until the Middle Ages.

In the Hellenistic period, Eratosthenes accurately measured the earth's circumference. Realizing that sunlight reached deep into a well at noon on the summer solstice in Syene near the Tropic of Cancer, he was able to determine the size of the earth. At noon on the summer solstice, a stick fixed vertically in the ground had no shadow in Syene, while another stick fixed in the ground in Alexandria had a shadow tilted

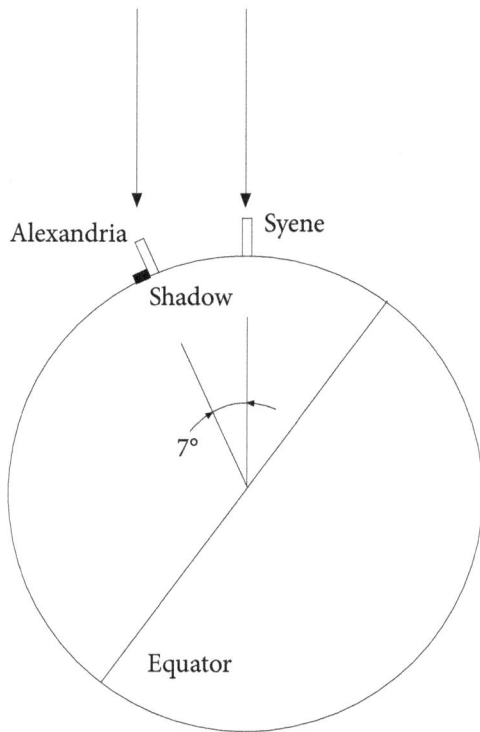

Figure 2-3 Eratosthenes' idea

32

seven degrees. Based on this discovery, he drew an image like Figure 2-3. It was already known that Syene is about 900 km from Alexandria. Thus, based on Figure 2-3, he estimated that the earth's circumference was 46,000 km. This result is very close to the actual circumference of the earth, 40,000 km.

Hipparchos, who also lived in the Hellenistic period, classified stars in terms of magnitude from 1 (brightest) to 6 (barely visible). This scheme is still employed today. For example, Venus and the full moon are minus 4 and minus 12 magnitude, respectively.

28 Constellations of the East

The Eastern tradition places seven constellations in each of the four cardinal directions (east, west, north and south), resulting in 28 constellations. The 28 constellations are marked by seven respective representative stars.

Gak, Hang, Jeo, Bang, Sim, Mi and Gi (각항저방심미기),
the blue dragon constellations represent the eastern sky.

Du, Woo, Yeo, Heo, Wi, Sil and Byeok (두우여허위실벽),
the black turtle constellations represent the northern sky.

Gyu, Roo, Wi, Myo, Pil, Ja and Sam (규루위묘필자삼),
the white tiger constellations represent the western sky.

Jeong, Gwi, Yu, Seong, Jang, Ik and Jin (정귀유성장익진),
the red phoenix constellations represent the southern sky.

Figure 2-4 shows the four guardian deities and the 28 constellations. The center of the image corresponds to the north pole of the sky, occupied by Polaris. Polaris and the famous Big Dipper do not belong to the 28 constellations. The respective seven constellations of east, west, south and north are located on the blue dragon, the white tiger, the red phoenix and the black turtle, respectively.

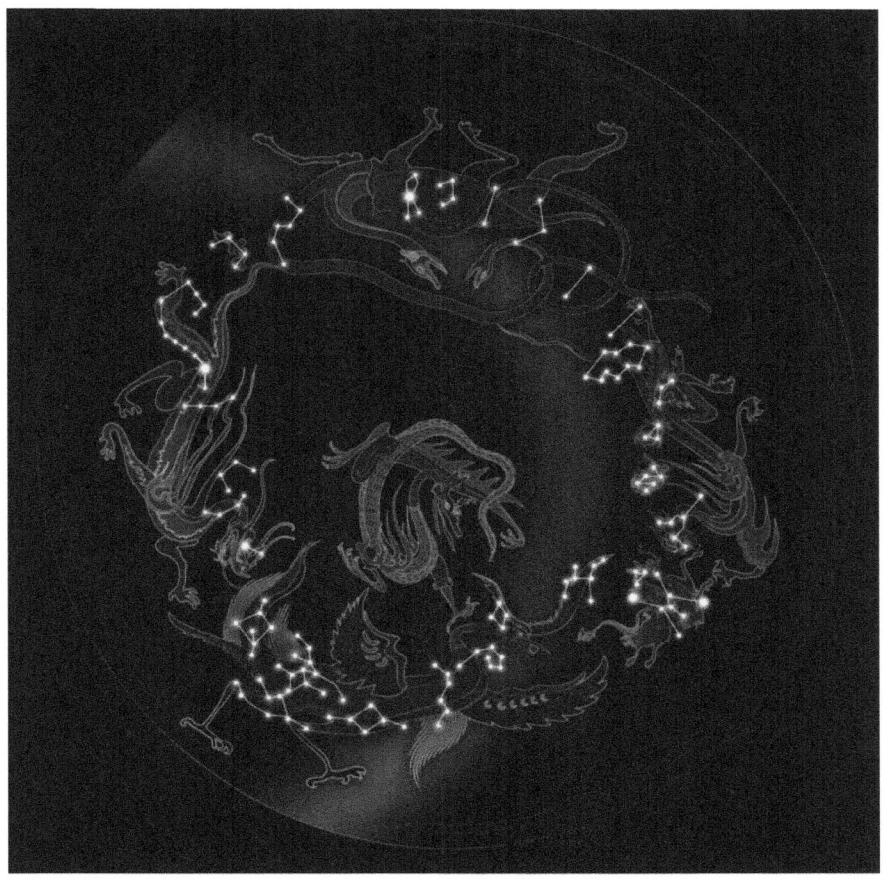

Figure 2-4 Four guardian deities (provided by Hong-Jin Yang)

Please note that in Figure 2-4, the blue dragon, white tiger, black turtle and red phoenix are on the left, right, top and bottom, respectively. East and west seem to be reversed compared to the four cardinal directions of maps. This is because the sky was drawn looking up from the floor.

You should also be aware that the eastern seven constellations, for example, do not indicate those appearing in the eastern sky all the time.

35

Any constellation in the eastern sky appears high in the southern sky and then finally sets in the west. We can see the eastern seven constellations on summer nights. Likewise, the seven northern, western and southern constellations are visible on autumn, winter and spring nights, respectively.

Order	English	Korean	Representative star	Order	English	Korean	Representative star
1	Gak	각	Virgo α	15	Gyu	규	Andromeda ζ
2	Hang	항	Virgo κ	16	Roo	루	Aries β
3	Jeo	저	Libra α	17	Wi	위	Aries 35
4	Bang	방	Scorpius π	18	Myo	묘	Taurus η
5	Sim	심	Scorpius σ	19	Pil	필	Taurus ε
6	Mi	미	Sagittarius μ	20	Ja	자	Orion λ
7	Gi	기	Sagittarius γ	21	Sam	삼	Orion δ
8	Du	두	Sagittarius φ	22	Jeong	정	Gemini μ
9	Woo	우	Capricornus β	23	Gwi	귀	Cancer θ
10	Yeo	여	Aquarius ε	24	Yu	유	Hydra δ
11	Heo	허	Aquarius β	25	Seong	성	Hydra α
12	Wi	위	Aquarius α	26	Jang	장	Hydra ν
13	Sil	실	Pegasus α	27	Ik	익	Crater α
14	Byeok	벽	Pegasus γ	28	Jin	진	Corvus γ

Table 2-2 The 28 constellations of the East

Table 2-2 presents the 28 constellations. In Table 2-2, the Greek letters α, β, γ, etc. represent the descending order of the stars' brightness.

For example, among the 28 constellations, Gak corresponds to Virgo α, meaning the brightest star in Virgo. Heo is represented by Aquarius β, meaning the second-brightest star in Aquarius. After all the 24 Greek letters are used, the subsequent stars can be numbered like Star No. 25, Star No. 26, etc.

2 zodiac constellations of the West

The planets do not move at random. The moon and the planets travel along a line, which astronomers call the ecliptic of heaven. The equator and the ecliptic are the most important lines in astronomy. As the solar system is a single plane, the planets are on the ecliptic. That is, the ecliptic is the plane of the solar system visible to us.

If you want to see a planet, you should find its ecliptic first. For example, since the Big Dipper is far from the ecliptic, it is useless to try to find a planet around it. As the ecliptic forms a great circle in the sky, it meets many constellations.

For this reason, if you remember those constellations, it will be of great help in studying planets. As the earth orbits around the sun in one year, the sun appears to make a complete circuit of the ecliptic over one year. Accordingly, by identifying the 12 constellations crossing the ecliptic, you can conveniently observe the course of the sun because it will meet one constellation in each month. From time immemorial, the zodiac signs of Table 2-3 were inherited in the Western tradition.

Month	English	Korean	Sign
1	Capricornus	염소	♑
2	Aquarius	물병	♒
3	Pisces	물고기	♓
4	Aries	양	♈
5	Taurus	황소	♉

Month	English	Korean	Sign
6	Gemini	쌍둥이	♊
7	Cancer	게	♋
8	Leo	사자	♌
9	Virgo	처녀	♍
10	Libra	천칭	♎
11	Scorpius	전갈	♏
12	Sagittarius	궁수	♐

Table 2-3 The 12 zodiac signs of the West.

Like the names of the planets, the signs and Latin names of the zodiac constellations go back thousands of years. The zodiac indicates to which constellation the sun belongs in each month. For example, according to Table 2-3, the sun is located in Gemini on a day in mid-June. That is, the sun and the earth are arranged as illustrated in Figure 2-5 so that the sun appears in the direction of Gemini from the perspective of the earth. Although the sun rises and sets due to the earth's rotation, the sun remains in the direction of Gemini. As Gemini rises and sets with the sun, it cannot be seen at night. This means that no one can see their zodiac sign on his or her birthday.

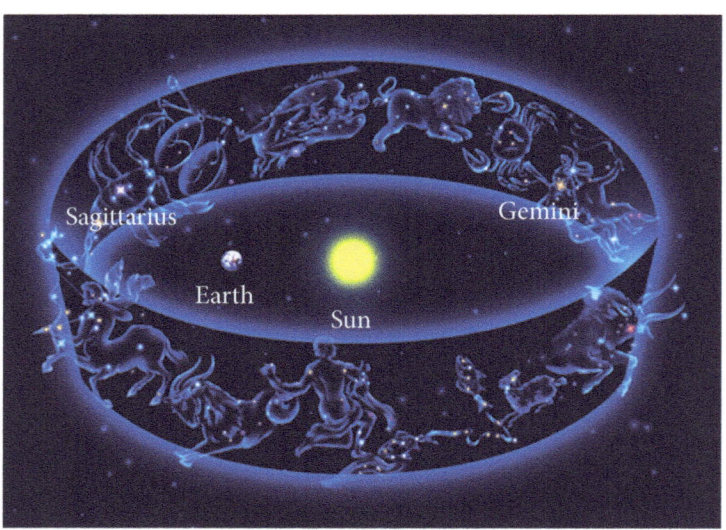

Figure 2-5 Image of 12 zodiac constellations
(created by Seon-Wook Kang)

After six months, that is, in mid-December, the earth has moved to the opposite side of the sun, and the sun has moved toward Sagittarius. Accordingly, Gemini is clearly seen on a mid-December night, meaning that one's zodiac sign becomes clearly visible in the sky six months after their birthday. Primitive people were intelligent enough to identify the constellations behind the sun shining brightly during the day.

The moon's path across the sky nearly coincides with the ecliptic, but there is an approximate 5-degree tilt between them.

Big Dipper and Southern Cross

Primitive people used the sun's position or the shadows of objects to determine the time of day. However, with nothing but constellations available at night, they mainly utilized the Big Dipper (북두칠성) in the northern hemisphere and the Southern Cross (남십자성) in the southern hemisphere.

As you see in Figure 2-6, the Big Dipper circles around Polaris counterclockwise once a day. If you watch it all night, you will recognize

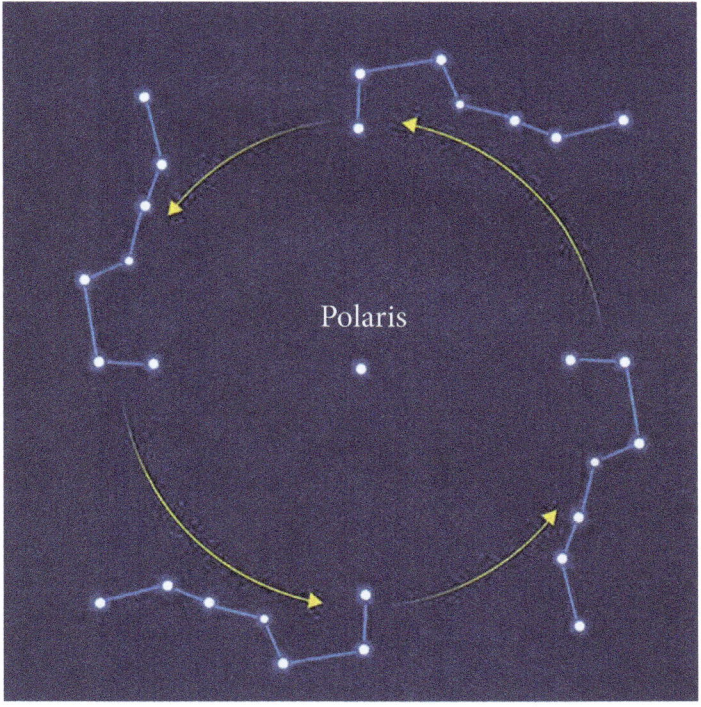

Figure 2-6 Big Dipper circling around Polaris once a day

three of the four shapes in Figure 2-6. We cannot see the remaining one because it appears only during the day. In spring, for example, the Big Dipper is on the right of Polaris in early evening and then rises over our heads as night approaches. In early morning, when the day breaks from the left of Polaris, the Big Dipper gradually disappears.

The Big Dipper consists of seven shining stars. In the current Western astrological pantheon, the Big Dipper corresponds to the tail of the Great Bear, which is a northern constellation. Hence, the Big Dipper itself is not an independent constellation but a part of Ursa Major. If we extend a line northward from the two stars at the end of the head part of the Big Dipper, the line will meet Polaris. These two stars are called the Pointers.

Since it serves as a means to find Polaris, the Big Dipper is treated as an indispensable tool even in army field manuals. We are lucky that the Big Dipper is close to Polaris. There is no star like it in the sky of the southern hemisphere. The star at the meeting point of the head and handle of the Big Dipper is fourth from either side. The remaining six stars are brighter. This is because the fourth star is a third-magnitude star, while the others are second-magnitude stars. Polaris is also a second-magnitude star.

The Big Dipper is inseparable from the Korean people. A folktale teaches that we are all born blessed with the god of the Big Dipper. The Big Dipper also accompanies a person at their death. In the traditional Korean funeral culture, the thin board laid on the bottom of a coffin is called the board of the Big Dipper (칠성판). The seven fairies participating in the ceremony of heaven's opening also symbolize this constellation.

I said before that the full moon rotates over time but cannot turn upside down. The same applies to the Big Dipper. While circling around Polaris as shown in Figure 2-6, it never has the shape in Figure 2-7.

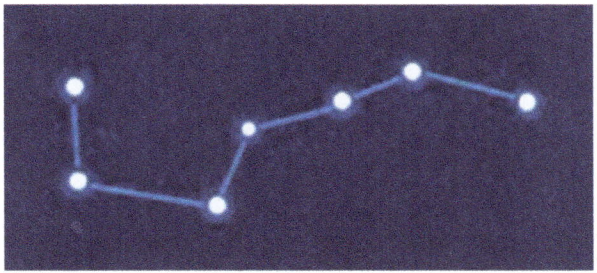

Figure 2-7 An incorrect upside-down image of the Big Dipper

Since there is no star like Polaris in the sky of the southern hemisphere, the Southern Cross revolving clockwise should be used like the Big Dipper. The Southern Cross consists of four shining stars. Its Latin name is "Crux," meaning cross. As shown in Figure 2-8, the longer line points to the south pole of the sky.

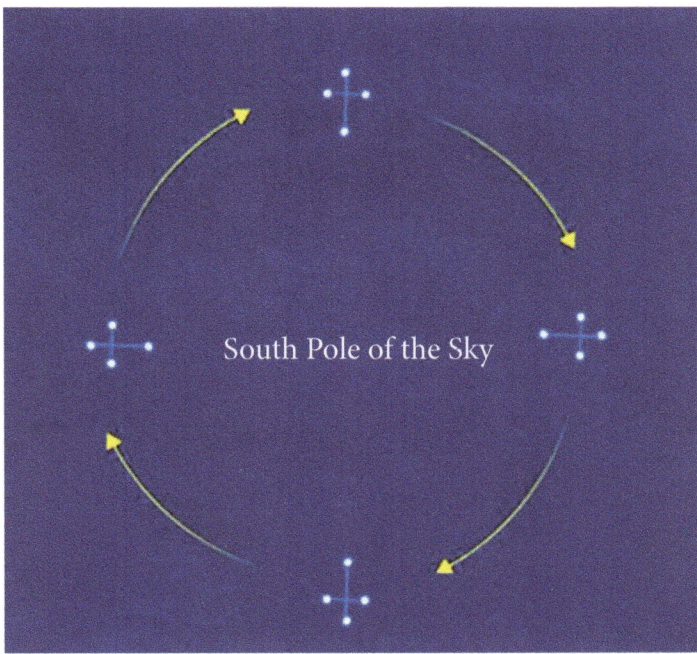

Figure 2-8 Southern Cross revolving around the south pole of the sky once a day

Cosmic God and His Disciples: Travel across the Universe

Cosmic god: "Can we speed it up a little?"
Galaxy god: "Row faster."
Earth god: 'Why is it always me?'

Cosmic god: "Comets never disappoint!"
Galaxy god: 'Gah, I'm scared to death!'
Earth god: 'I'm about to fall off.'

Chapter 3

Oriental Cosmos Worked with Taegeuk

The cosmology of the East can be summarized by Cheonbugyeong, Hado, Bokhui's Eight Trigrams, Nakseo and King Wen's Eight Trigrams. Cheonbugyeong was passed down orally from time immemorial, and Bokhui invented Hado and the Eight Trigrams in the era of Hwanung Baedal. During the Dangun Joseon period, King Yu of Xia dynasty and King Wen of Zhou dynasty made Nakseo and the other Eight Trigrams, respectively. Ancient Eastern cosmology was more sophisticated than its Western counterpart.

Cheonbugyeong

As the posterity of heaven, the Korean people have some written forms of cosmology, not mythology. Cheonbugyeong (천부경), The Scripture of Heavenly Code, is one of the representative cosmological texts.

1	시	무	시	1	석	3	극	무
진	본	천	1	1	지	1	2	인
1	3	1	적	10	거	무	궤	화
3	천	2	3	지	2	3	인	2
3	대	3	합	6	생	7	8	9
운	3	4	성	환	5	7	1	묘
연	만	왕	만	래	용	변	부	동
본	본	심	본	태	양	앙	명	인
중	천	지	1	1	종	무	종	1

This scripture is very difficult to understand. Numerous interpretations have been suggested but they are all different. As 31 out of 81 characters are numbers, you must be familiar with numbers to understand the meaning of Cheonbugyeong.

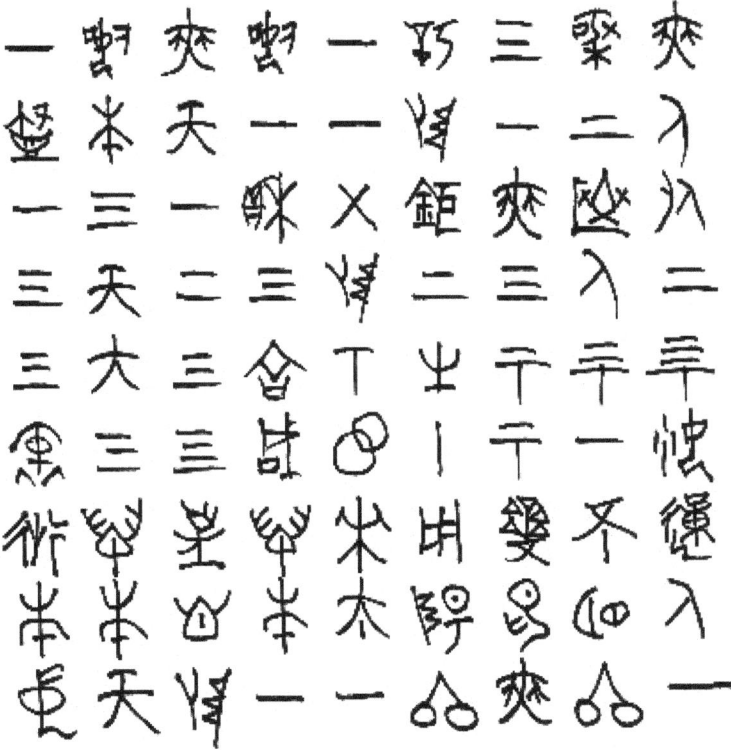

Figure 3-1 Cheongbugyeong in shell-and-bone characters
(excerpt from Gaecheongi)

According to the cosmology of yin and yang, odd numbers are heavenly numbers (천수) with yang energy, while even numbers are earthly ones (지수) with yin energy. The odd number 1 is the number of Taegeuk (태극). Since Taegeuk is the state before the force of the universe is divided into yin and yang, 1 symbolizes the beginning and principle of the universe. In other words, all numbers originate from 1. In the western tradition, Thales argued that water was the principle of everything. In the same vein, 1 symbolizes water (수). It also represents north, which is the fundamental place of all change. This corresponds to winter.

The earthly number 2 is divided into yin and yang. Although originating from 1, 2 becomes a parent on par with 1 and produces all the other numbers. Accordingly, 1 represents heaven and 2 the earth. In addition, where 1 represents water, 2 represents fire. This calls to mind Heraclitus, who picked out fire as the fundamental substance of the world. In addition to symbolizing fire (화), 2 signifies the opposite direction of north, that is, the warm south. Naturally, the corresponding season is summer.

The heavenly 3 is the first number made by combining the heavenly number 1 and the earthly number 2. Being the first number produced by the combination of yin and yang, 3 symbolizes spring, when everything bursts into life. It is natural that 3 represents wood (목) and east, where the sun rises. From an abstract perspective, 3 is the smallest number capable of constituting the universe. 3 provides the ground for the argument that the universe is the harmony of heaven, earth and man. That is, 3 is like the synthesis in the thesis-antithesis-synthesis paradigm.

The earthly number 4 represents metal (금), which is also a constituent of the universe together with water, fire and wood. 4 corresponds to the remaining season, autumn. West is its direction. When metal is added, the universe is completely configured, like the four directions (east, west, south and north) and the four seasons (spring, summer, autumn and winter). This can be compared with the Aristotelian universe made of four elements, water, fire, air and earth.

The heavenly number 5 is the agent of harmony between yin and yang. 5 represents earth (토). The other elements, water, fire, wood and metal, which are symbolized by 1, 2, 3 and 4 respectively, cannot exist apart from earth. Thus, being the base of everything under the sky, earth occupies not a specific direction but the center.

In this way, the ancient eastern view of the universe was completed by the two energies (yin and yang) and the five elements (wood, fire, earth, metal and water). This is also called the cosmology of yin, yang and the five elements (음양오행). The order of wood, fire, earth, metal and water corresponds to the order of mutual life-giving or spring, summer, autumn and winter.

Bokhui categorized 1, 2, 3, 4 and 5 as generation numbers and 6, 7, 8, 9 and 10 as formation numbers. The generation numbers (생수) generate all things, while the formation numbers (성수) form them. Based on this understanding, let's read *Cheonbugyeong (The Scripture of Heavenly Code)*. The Daehan History and Culture Association provides the following English translation.

Cheonbugyeong

One is the beginning; from Nothingness begins One.

One divides into the Three Ultimates, yet its substance remains inexaustible.

Arising from One, Heaven is One.

Arising from One, Earth is Two.

Arising from One, Humanity is Three.

One accumulates and climaxes at Ten, as Three governs the change of all things.

Based on Two, Heaven changes under Three.

Based on Two, Earth changes under Three.

Based on Two, Humanity changes under Three.

The great Three unite into Six, which then gives rise to Seven, Eight, and Nine.

Everything moves in accordance with Three and Four; everything circulates under Five and Seven.

Our proliferates in mysterious ways, evolving in perpetual cycles, and these functions ultimately transform into immutable substance.

The ultimate substance is Mind, which shines radiantly like the sun.

Humanity, united with Heaven and Earth, is the Ultimate One.

One is the end; in Nothingness ends One.

Four signs and eight trigrams

Bokhui's universe of yin, yang and five elements began to evolve as Taegeuk was divided into yin and yang. The universe we live in is made of two energies: yin and yang. Yang is represented by the following solid line:

Yin is represented by the following broken line:

The above two lines are called yangui (양의). When yangui is divided, four signs (사상) are obtained. The four signs are the different possible pairs of the yin and yang lines. These pairs are yang-yang, yin-yin, yang-yin and yin-yang.

① The yang-yang pair is called tae yang (태양), full of yang energy.

② The yin-yin pair is called tae yin (태음), full of yin energy.

③ The yang-yin pair is called so yang (소양), little yang.

④ The yin-yang is called so yin (소음), little yin.

From the four signs, the following eight trigrams (팔괘) are generated.

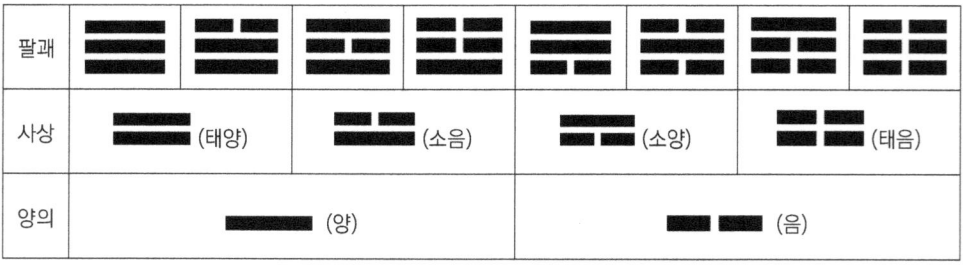

Table 3-1 Generation of Eight Trigrams

Bokhui explained each trigram as follows:

① Geon (건)

This trigram represents the heaven. Consisting of three yang lines, this trigram is nothing but the heaven.

② Tae (태)

This represents a pond or the sea. The yang lines blocking from below create an image of pooling water.

③ Li (이)

This represents fire or sun. Two yang lines surround one yin line; it is bright outside but dark inside.

④ Jin (진)

This represents thunder or lightning. The force of the yang line at the bottom seems to soar through the two yin lines. In the past, people believed that lightning flashed upward through the ground.

⑤ Son (손)

This trigram represents wind. The force of yang lines comes down to the yin line.

⑥ Gam (감)

This represents water or the moon. As the yang line is surrounded by two yin lines, it is bright inside but dark outside, like water.

⑦ Gan (간)

This represents the mountain. One yang line is on two yin lines. Yang can rise no further. This symbolizes the shape of a mountain ridge.

⑧ Gon (곤)

This represents earth. Consisting of three yin lines, this trigram is nothing but earth.

You can memorize the order of the eight trigrams as follows: 1 Geon, 2 Tae, 3 Li, 4 Jin, 5 Son, 6 Gam, 7 Gan and 8 Gon. Table 3-2 presents the eight trigrams.

Order	Trigram	English	Korean	Meaning
1	☰	Geon	건	Heaven
2	☱	Tae	태	Pond, sea
3	☲	Li	이	Fire, sun
4	☳	Jin	진	Thunder, lightning
5	☴	Son	손	Wind
6	☵	Gam	감	Water, moon
7	☶	Gan	간	Mountain
8	☷	Gon	곤	Earth

Table 3-2 Bokhui's eight trigrams

Bokhui's eight trigrams

Taeho Bokhui's genius is illustrated by the circular arrangement of the Eight Trigrams as seen below in Figure 3-2. In that figure, 1 Geon, or heaven, is on top, while 8 Gon, or the earth, is on the bottom. Heaven and earth become the frame. In Bokhui's Eight Trigrams, the sum of the numbers indicated by any two opposing trigrams is always 9. For example, the sum of 3 Li and 6 Gam is 9. Any pair of opposite trigrams have shapes symmetrical with each other. Moreover, if the yang line is one piece and the yin line consists of two pieces, any two opposite trigrams include nine pieces. For example, since 3 Li has four pieces and 6 Gam does five pieces, there are nine pieces in total. How marvelous it is!

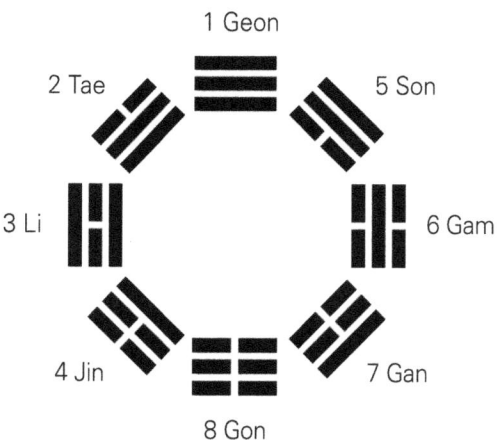

Figure 3-2 Bokhui's Eight Trigrams

Using the binary system of 0 and 1, eight permutations with repetition can express a three-digit number. These cases are 000, 001, 010, 011, 100, 101 and 111. If 0 and 1 are replaced by the yin line and the yang line respectively, we can obtain Bokhui's Eight Trigrams.

On the left side of 1 Geon come 2 Tae, 3 Li and 4 Jin. On the right side of 1 Geon, 5 Son, 6 Gam and 7 Gan are arranged in descending order. 8 Gon is located at the bottom. You should note that yang is the innermost line in Geon, Tae, Li and Jin, while yin is the innermost line in Son, Gam, Gan and Gon.

The upper part is warm, with more yang lines, corresponding to the south, while the lower part is cold with more yin lines, representing the north. Of course, this is because people sit facing south in the sun. Accordingly, the left side of the Eight Trigrams is east, and the right side is west. The trigram of Li is in the east and Gam is in the west. This is because the full moon is in the western sky while the sun is in the eastern sky. Thus, the four heavenly, astronomical trigrams — Li, Gam, Geon and Gon — are located in the east, west, south and north, respectively. It is natural that the directions are determined based on the sun and the moon.

In Figure 3-3, a folding screen depicting the sun in the east and the moon in the west stands behind the throne of Geunjeongjeon (근정전) in Gyeongbokgung (경복궁) palace, Seoul. It means the sun, the moon and five mountain peaks. That is, the folding screen represents the universe of yin, yang and five elements (음양오행). It perfectly fits the ruler of the heavenly descendants.

Figure 3-3 Throne of Geunjeongjeon of Gyeongbokgung palace and the folding screen of sun, moon and five mountain peaks

Now, let's examine the arrangement of the four earthly, geographical trigrams: Tae, Jin, Son and Gan. Bokhui lived on the Chinese mainland because it was a part of Korea at that time. Accordingly, the trigram of Gan is in the northwest because there are mountains in that direction. As mountains rise from the ground, it is natural that the trigram of Gan is next to Gon. The trigram of Tae is placed in the southeast, since the sea is there. As lightning soars up from the ground, the trigram of Jin is next to Gon. As the wind descends from the sky, Son is located next to Geon.

Thus, as shown in Figure 3-2, the four heavenly trigrams are placed in the four cardinal directions (east, west, south and north), while the four earthly trigrams take the diagonal directions. If the Eight Trigrams thus arranged are turned upside down, the heavenly trigrams do not shape, but the earthly ones do. The mastery of heavenly and earthly principles means thoroughly understanding the principle of the Eight Trigrams.

The shape of Taegeuk can be derived from Bokhui's Eight Trigrams. In Figure 3-4, the circle surrounded by the trigrams is divided into eight sectors. The sector right below the trigram of Geon is completely white. As the trigram of Geon has no yin line but three yang lines, the sector is filled with white, that is, the force of yang. On the contrary, the sector right over the trigram of Gon is completely gray. Since the trigram of Gon consists of three yin lines and no yang line, the corresponding sector is colored gray, representing the energy of yin.

As for the sectors corresponding to the trigrams of Tae and Son, two-thirds are colored white and one-third is colored gray. This is because both trigrams have two yang lines and one yin line. Likewise, in the sectors corresponding to the trigrams of Jin and Gan, two-thirds of the area is gray and the remaining one-third is white.

Figure 3-4 Archetype of Taegeukgi

Based on the same principle, two-thirds of the sector for the trigram of Li should be colored white and the remaining one-third should be gray, since the trigram consists of two yang lines and one yin line. However, the sector is half white and half gray. This may be because the one yin line is in the middle of the two yang lines. Likewise, the sector corresponding to the trigram of Gam is half white and half gray. The shape of Taegeuk is naturally derived by connecting the lines.

In Bokhui's Eight Trigrams, those facing each other are in equilibrium and harmony. Thus, Bokhui's Eight Trigrams display the ideal universe. Unfortunately, this is just an ideal and far from reality. We know that neither equilibrium nor harmony is present in reality. For this reason, King Wen suggested his eight trigrams reflecting such a reality.

Hado

Taeho Bokhui drew the image of Hado to describe the cosmic cycle based on Cheonbugyeong. Its Korean name (하도) means an image illustrating the flow of energy. According to the legend, when a dragon horse with Hado on its back appeared from a river named Hasoo, Bokhui drew Hado. As shown in Figure 3-5, Hado consists of white and black points.

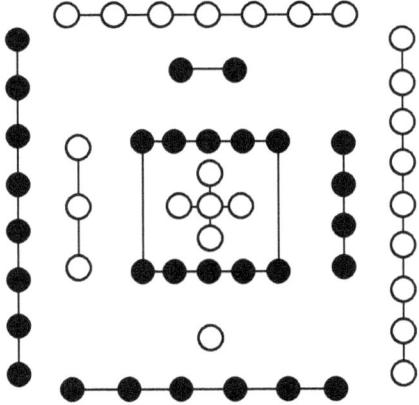

Figure 3-5 Hado

In Hado, the last generation number — 5 — and the last formation number — 10 — are placed at the center. The generation numbers 1, 2, 3 and 4 are placed in the inner part, while the formation numbers 6, 7, 8 and 9 are in the outer part. As mentioned, the odd numbers 1, 3, 5, 7 and 9 are heavenly numbers, and the even numbers 2, 4, 6, 8 and 10 are earthly numbers. Naturally, the heavenly numbers represent yang energy, while the earthly ones represent yin energy. As the heavenly and earthly numbers are indicated by white and black points, respectively, Hado can be shown through the Badook board in Figure 3-6.

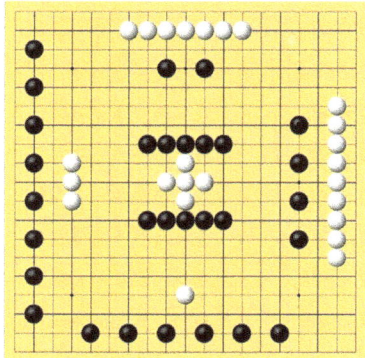

Figure 3-6 Hado on a Badook board

The number 5 is at the center surrounded by the heavenly numbers 1, 3, 7 and 9. Thus, 5 takes the central position again in Hado. Once five white points are placed at the center, the remaining heavenly numbers are 1, 3, 7 and 9. Accordingly, Bokhui arranged 1, 3, 7 and 9 along each side. As the generation numbers should occupy an inner location and the formation ones an outer location, 1 and 3 are placed in the inner part and 7 and 9 are located in the outer part, as shown in Figure 3-7.

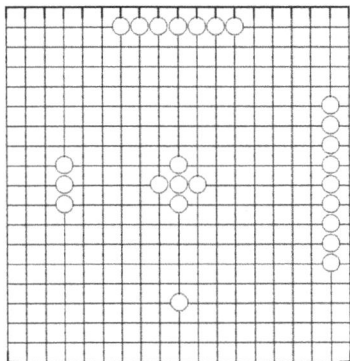

Figure 3-7 White stones of Hado on a Badook board

Here, the number 1 placed at the bottom indicates the north, as in Bokhui's Eight Trigrams. The sum of 5 (at the center) and 1 is 6. Accordingly, as illustrated in Figure 3-8(A), 6 is placed at the bottom. Since 6 is an earthly number, it should be indicated by black stones, and as it is a formation number, it should be at an outer place. Thus, the heavenly number 1 and the earthly number 6 form a pair placed at the bottom of Hado. When the generation number 1 conceives water, the formation number 6 constitutes it.

Similarly, the addition of 3 to the center (5) makes 8. Accordingly, as shown in Figure 3-8(B), 8 is on the left side of 3. As 8 is an earthly number, it should be expressed by black stones, and being a formation number, it should be in the outer part. Thus, the pair comprising the heavenly number 3 and the earthly number 8 is located on the left. The generation number 3 brings forth fire, and the formation number 8 completes it.

Subtracting the center (5) from 7 gives us 2. Since it is a formation number, 7 is already located in the outer part. Accordingly, as shown in Figure 3-8(C), 2 is in the inner part relative to 7. Being an earthly num-

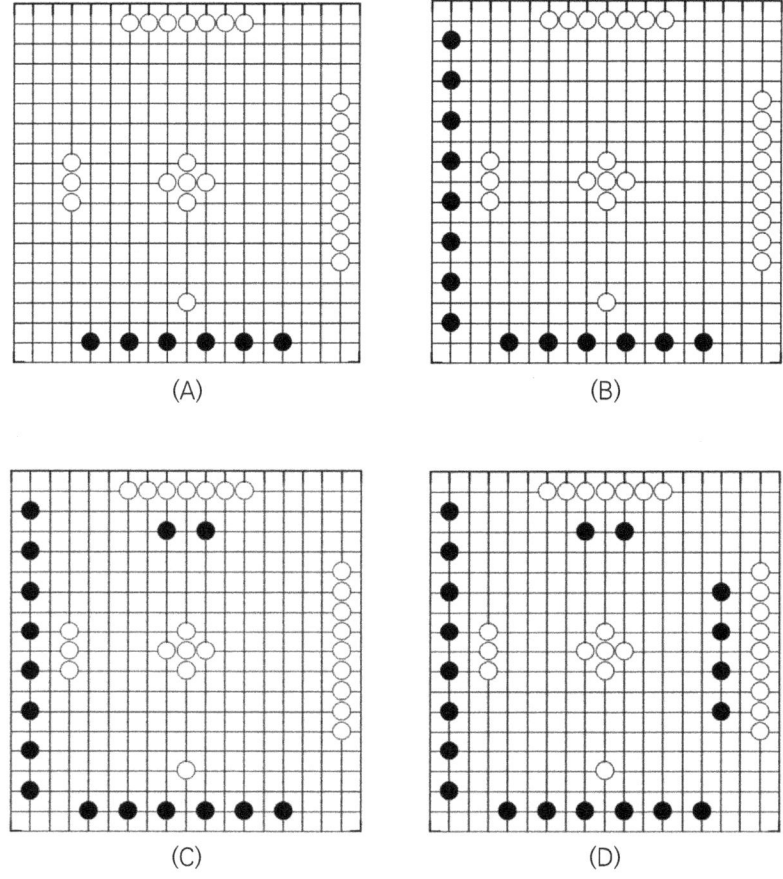

Figure 3-8 Arrangement of black stones in Hado on a Badook board

ber, 2 should be expressed by black stones. Being a generation number, it should belong to the inner part. Thus, the pair of the earthly number 2 and the heavenly number 7 is located on the top. While 2 as a generation number brings forth wood, 7 as a formation number completes it.

Finally, the subtraction of the center 5 from 9 makes 4. Being a formation number, 9 is already located in the outer part. Accordingly, as shown in Figure 3-8(D), 4 is further inside than 9. As an earthly generation number, 4 should be expressed by black stones and located in the inner section. Thus, the pair comprising the earthly number 4 and the heavenly number 9 is located on the right side of Hado. 4 as a generation number brings forth metal, and 9 as a formation number completes it.

Now, if you add 10 black stones to Figure 3-8(D), the complete Hado of Figure 3-6 is obtained. When the generation number 5 produces earth, the formation number 10 completes it. As a result, 55 stones are needed to complete Hado. Here, the number of white stones is 25 (1+3+5+7+9), and that of black stones is 30 (2+4+6+8+10). If the central numbers 5 and 10 are excluded, then the sum of the heavenly numbers is 20 (1+3+7+9), and that of the earthly numbers is also 20 (2+4+6+8).

In Figure 3-6, the line connecting the white stones 1 → 3 → 7 → 9 indicates that the force of yang moves out from north to south in a clockwise swirl. The line connecting the black stones 8 → 6 → 4 → 2 indicates that the force of yin moves in from south to north in a counterclockwise swirl. This also makes the shape of Taegeuk.

In every direction of Hado, we find a pair made by a generation number of yang and a formation number of yin, or a generation number of yin and a formation number of yang.

Now, let us apply the principle of five elements to Figure 3-9. Imagine a clockwise swirl moving from north below in the order of 1·6 (water) → 3·8 (wood) → 2·7 (fire) → 4·9 (metal). As water gives life to wood and wood does to fire, there is no problem until 1·6 → 3·8 → 2·7. However, since fire overcomes metal, the process 2·7 → 4·9 is not possible. Accordingly, it is necessary to proceed from 2·7 to 4·9 via the center 5·10. That is, the cycle of mutual life-giving can be made in the form of 2·7 (fire) → 5·10 (earth) → 4·9 (metal). This process is not problematic since fire gives life to earth, as does earth to metal and metal to water.

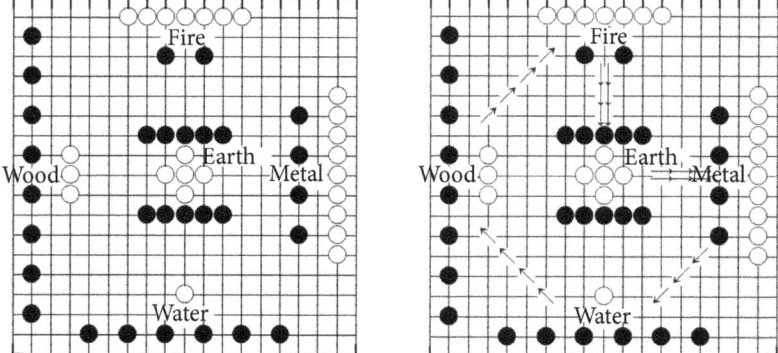

Figure 3-9 Life-giving cycle of Hado

Nakseo and King Wen's eight trigrams

King Yu of Xia dynasty made Nakseo (낙서) to explain the cosmic cycle. Legend tells that King Wen copied Nakseo on the back of a turtle appearing from the Nakseo, which also consists of white and black points. Hado illustrates the cosmic structure, while Nakseo shows the cosmic evolution.

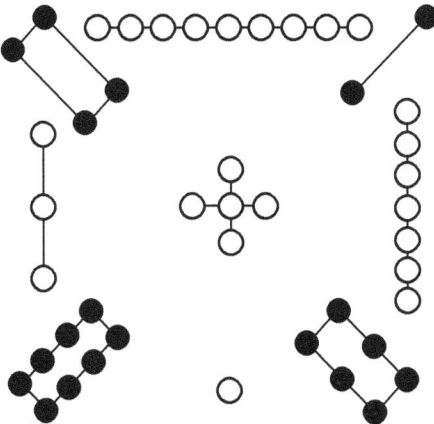

Figure 3-10 Nakseo

Hado discloses the substance of the universe by means of the numbers 1 to 10, while Nakseo explains the use of universe using the numbers 1 to 9. The number 5 is like a fetus in a mother's womb, that is, 10 in Hado. However, 5 appears to be independent in Nakseo. Figure 3-11 shows Nakseo arranged on a Badook board.

Unlike Hado, Nakseo places the heavenly numbers 1, 3, 7 and 9 in the north, east, west and south, respectively, and the earthly numbers 2, 4, 6 and 9 in the diagonal directions. The number 5 at the center coordinates the other numbers.

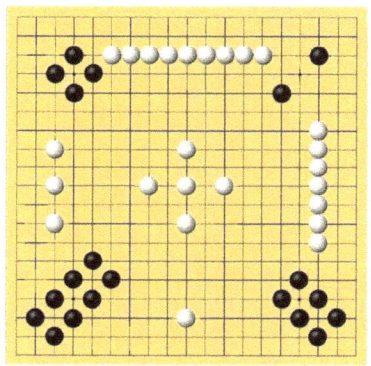

Figure 3-11 Nakseo on Badook board

The force of yang expands in the order of 1 → 3 → 9 → 7 in a clockwise swirl, while the force of yin contracts in the order of 8 → 6 → 2 → 4 in a counterclockwise swirl. Hado mixes yin and yang in every direction, but Nakseo divides them.

In Hado, with 5 at the center, the sum of any two opposite numbers facing each other is 10. In other words, 10 is everywhere in Nakseo. This is possible only when 5 takes the center position. Thus, we obtain a magical square in which the sums of three numbers in each row, each column, and both main diagonals are 15.

Also, in Nakseo, 1 and 6 correspond to water, 2 and 7 to fire, 3 and 8 to wood, 4 and 9 to metal, and 5 and 10 to earth. For this reason, Bokhui's five elements are arranged as in Figure 3-12. If you compare Figure 3-12 with Figure 3-9, you can see that metal and fire swap positions.

Nakseo shows a cycle in a counterclockwise swirl in the order of water, fire, metal, wood and earth, which is opposite to that of Hado. In other words, Nakseo displays the cycle of mutual conflict: fire overcome by water → metal overcome by fire → wood overcome by metal → earth overcome by wood → water overcome by earth.

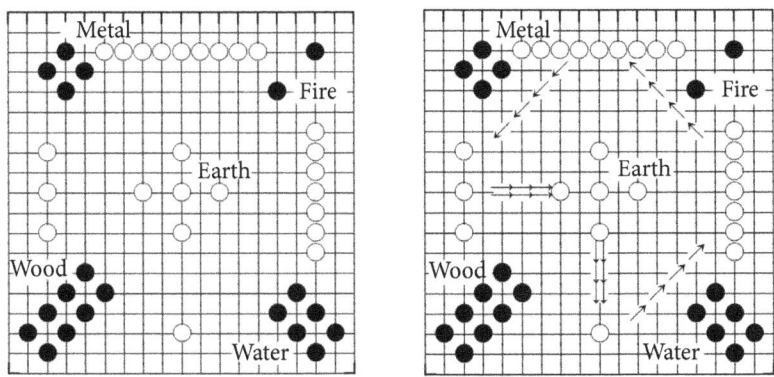

Figure 3-12 Cycle of mutual conflict in Nakseo

I have explained that Hado and Nakseo illustrate the cosmic structure and the cosmic evolution, respectively. Hado corresponds to the substance, and Nakseo to the use. In both Hado and Nakseo, yang expands in a counterclockwise swirl and yin contracts in a clockwise swirl. In Hado, 10, at the center, controls yin and yang. However, in Nakseo,

10 has no such control. Moreover, the movements of Hado form a harmony between yin and yang, while those of Nakseo are independent.

King Wen of Zhou proposed eight trigrams that are essentially the same as Nakseo. They are called King Wen's eight trigrams. King Wen was one of the dukes of Dangun Joseon, the founders of the Zhou dynasty. Bokhui's eight trigrams show the equilibrium and harmony between each pair of trigrams that face each other. That is, Bokhui's eight trigrams represent the ideal figure of the universe.

However, it seems quite removed from reality because no equilibrium or harmony is found in the real world. Mutual conflict appears to dominate over mutual life-giving. King Wen's eight trigrams tried to reflect such a reality and can be illustrated as in Figure 3-13.

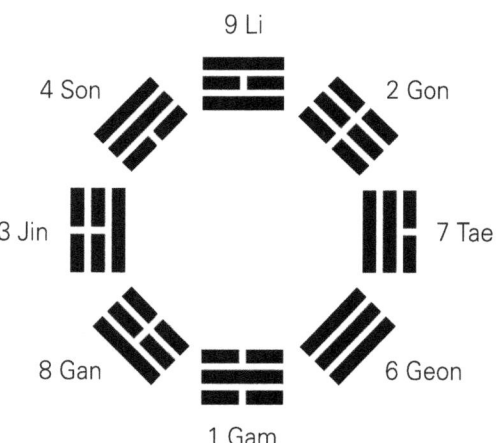

Figure 3-13 King Wen's eight trigrams

In Bokhui's eight trigrams, the Geon trigrams and the Gon trigrams, which are like the parents of the universe, grasp the central axis on the top and bottom, respectively. In King Wen's eight trigrams, the Gam trigram and the Li trigram play a comparable role. The numerical arrangement of King Wen's eight trigrams is as follows:

$$\begin{array}{ccc} 4 & 9 & 2 \\ 3 & & 7 \\ 8 & 1 & 6 \end{array}$$

You may notice that the above arrangement can be obtained by removing the center number 5 from Nakseo. The sum of any two numbers facing each other is exactly 10. For example, the sum of 3 Jin and 7 Tae is 10. Thus, it is obvious that King Wen's eight trigrams are related to Nakseo. In Bokhui's eight trigrams, the sum of any two numbers on opposite sides is 9.

King Wen's eight trigrams are arranged in the following order: 1 Gam, 2 Gon, 3 Jin, 4 Son, 6 Geon, 7 Tae, 8 Gan, and 9 Li. Compare them with Bokhui's eight trigrams: 9 is added while 5 is removed. Some scholars point out a relationship between Bokhui's eight trigrams and Hado. This may be because Bokhui made both. However, there is no evidence as clear as that for the relationship between King Wen's eight trigrams and Nakseo.

The interpretation of King Wen's eight trigrams can be summarized as follows: Jin trigram has one yang line soaring up through two yin lines and represents the sprouting season of spring, and east. Son trigram symbolizes the growth of living things and thus represents the time between spring and summer, and southeast. Li trigram is so bright that all living things compete to grow, thereby representing summer, and south. Gon trigram is the earth, symbolizing the time between summer and autumn during which all living things flourish and are about to be harvested. Gon trigram indicates southwest. Tae trigram absorbs energy from below and bears fruit. It symbolizes the season of autumn, and west. Geon trigram represents the time between autumn and winter in which the cold energy becomes stronger and a fight against yin energy begins. Geon trigram symbolizes northwest. Gam trigram is water, representing winter and north. Gan trigram symbolizes northeast and the time between winter and spring in which all living things are dead and born again.

To sum up, all things are born in Jin trigram, settle in Son trigram, become bright in Li trigram, grow in Gon trigram, are harvested in Tae trigram, fight in Geon trigram, rest in Gam trigram and end in Gan trigram. Accordingly, in King Wen's eight trigrams, Jin in the east, Li in the south, Tae in the west and Gam in the north symbolize spring, summer, autumn and winter. In addition, Son in the southeast symbolizes the time between spring and summer, Gon in the southwest represents the time between summer and autumn, Geon in the northwest symbolizes the time between autumn and winter, and Gan in the northeast represents the time between winter and spring.

King Wen's eight trigrams form a clockwise swirl in the order of Jin → Son → Li → Gon → Tae → Geon → Gam → Gan. This follows the flow of spring, summer, autumn and winter. Now, we can see that Bokhui's eight trigrams are a spatial arrangement, while King Wen's eight trigrams are structured in terms of time.

Cosmic God and His Disciples: Pleasant Earth

Cosmic god: "I feel out of it after a hundred years' meditation. Let's play soccer."
Galaxy god: "Zzz…"
Earth god: 'Gosh, I was bored to death.'

Cosmic god: "How many light years for a penalty?"
Galaxy god: "That seems a bit far…"

Chapter 4

Our Heaven Opened with Gaecheon

Astronomical records prove that Hwanung Baedal and Dangun Joseon were not mythical countries. Now we have to establish the historical succession of Korean people before the Common Era, from Hwanin Hwankuk to Hwanung Baedal to Dangun Joseon and to Northern Buyeo. We should also realize that the National Foundation Day celebrates not Dangun but Hwanung. Gaecheon (Heavenly Opening) is referred to as the event where Hwanung performed works of benevolence with the herd of heavenly descendants. The spirit of the Baedal people is conveyed by the ideas of Gaecheon, heavenly posterity, and Hongik (benefit for all).

Dangun-Joseon as history, not myth

"Mujin Osipnyeon Oseong Chuiroo (무진오십년오성취루)"

This sentence in *Dangunsegi* (단군세기) is the record that the alignment of the five major planets was observed in the period of Dangun Joseon. Here, "Mujin Osipnyeon" refers to the year 1733 BCE; "Oseong" means the five planets Mercury, Venus, Mars, Jupiter and Saturn; "Chui" is "to collect, gather or come together" and "Roo" is one of the 28 constellations. Accordingly, the statement can be interpreted to mean that the five planets gathered beside the star called Roo in 1733 BCE.

Figure 4-1 Simulation of the alignment of the five planets
(created by Soo Jin Park)

Dangun Joseon had an observatory called Gamseong (감성). The astronomical record itself has significant meaning since it tells us that our ancestors already had an established culture of recording astronomical phenomena. Dangun Joseon was an ancient country, but it was also so advanced as to have an observatory.

Using astronomical software, we can confirm that the five planets — Mars, Mercury, Saturn, Jupiter and Venus — were aligned from left to right in the western sky on a mid-July evening in 1734 BCE, as illustrated in Figure 4-1. The five planets joined in a line, like beads on a string, between the sun and the moon in the evening of July 13, 1734 BCE.

Despite the difference of one year, it is reasonable for us to think the latter event is the actual alignment of five planets, since we are estimating events dating back as many as about 3800 years. As we have no means to guess the calendar of that time, our estimation seems to be still more persuasive. It is almost impossible to correctly guess or manually calculate astronomical events like planetary alignment without a computer.

Dangun Joseon was a great country, dominating the eastern part of the Eurasian continent. The three-territory system (삼한관경제) refers to the governance of Dangun Joseon, in which its large territory was divided into Mahan (마한), Beonhan (번한) and Jinhan (진한), as shown in Figure 4-2. These three states are collectively called Daehan (대한). Thus, Daehanminkuk (대한민국), another name for the Republic of Korea, literally means the country of Dangun Joseon.

Figure 4-2 Three Han states of Dangun Joseon (Excerpt from *Yuwija*)

Jinhan was governed by Dangun, while Beonhan and Mahan were governed by Vice-Dangun. Wanggeom, the first Dangun, sent the crown prince Buru to help Yu Sagong of the Yu dynasty prevent the nine-year flood. With this achievement, Yu Sagong founded the Xia dynasty and became the King Yu. As the Xia dynasty is the beginning of the Yellow River Civilization, the Chinese people call themselves the people of Huaxia. The Xia dynasty was followed by the Shang dynasty and the Zhou dynasty, both of which were peripheral vassal states of Dangun Joseon. As shown in Figure 4-2, these states did not occupy a large area. As all the feudal lords were Baedal people, it is not an exaggeration to say that the Huaxia people have no ancient history.

The king of Qin dynasty called himself the first emperor and denied the history of the Baedal people. The "burning of books and burying of scholars" was an attempt to remove every record testifying the dominance of the Baedal people. In other words, the king declared the new beginning of the Huaxia people, liberated from the seemingly everlasting governance of the Baedal people. With this goal in mind, he made his subjects build the Great Wall and called the peoples living north of the border barbarians. Nevertheless, his efforts have come to nothing because the historic sites of the Hong San culture were all found outside the Great Wall.

Lineage of Korean Sovereignty in BCE

The great history of the Baedal people began to decline after the collapse of Goguryeo (고구려). Many refugees were forced to migrate to neighboring countries, and some had no choice but to live their lives within the Korean peninsula. The founder of Joseon (not Dangun Joseon) thought that a small country's invasion into a larger one was violating Heaven's will, and China became something of an elder brother to Joseon. Joseon showed a deferential attitude to China by collecting and destroying history books stating that the continent was governed by

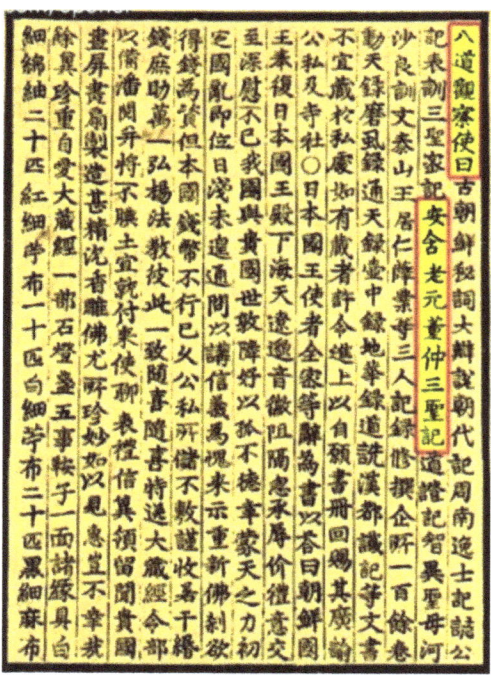

Figure 4-3 'Samsunggi' by Ham-Ro Ahn and Dong-Jung Won that was included in the King's order to collect history books

the Korean people. Some kings even killed those who possessed such texts. Only a list of lost books is left to us. For example, Figure 4-3 is a page of King Sejo's command to the provincial governors. The list includes *Samsunggi* (삼성기), written by Ham-Ro Ahn and Dong-Jung Won.

Terauchi Masatake, the first governor during the Japanese occupation, collected history books and spread colonial history as soon as he took office in Korea. As a result, most of the books recording the history of our ancestors dominating the continent were lost. In 1911, Yeon-Su Gye, an independence activist, collected the following five history books that had survived until then and integrated them into a single book named *Hwandangogi* (환단고기). Yeon-Su Gye published 30 copies of the book.

Samsunggi volume 1 (삼성기 상) by Ham-Ro Ahn
Samsunggi volume 2 (삼성기 하) by Dong-Jung Won
Dangunsegi (단군세기) by Am Lee
Bukbuyeogi (북부여기) by Jang Beom
Taebaekilsa (태백일사) by Maek Lee

Dangunsegi records the alignment of the five planets that was introduced above. The astronomical verification of the five planets' alignment also clearly proves that *Hwandangogi* is not a lie. *Hwandangogi* means the ancient record of Hwanin (환인), Hwangung (환웅) and Dangun (단군). These three sacred originators are called Samsung (삼성) and the corresponding period is Samsungjo (삼성조시대). *Hwandangogi* divides the period of the three sacred originators as follows:

Portraits of Hwanin, Hwanung and Dangun
in Samsung temple at Guwol mountain

① Hwanin Hwanguk (환국): reign by 7 Hwanins for 3,301 years between 7197 BCE and 3897 BCE
② Hwanung Baedal (배달): reign by 18 Hwanungs for 1,565 years between 3897 BCE and 2333 BCE
③ Dangun Gojoseon (조선): reign by 47 Danguns for 2,096 years between 2333 BCE and 238 BCE

Hwandangogi includes *Bukbuyeogi*, written by Jang Beom. Bukbuyeo or Northern Buyeo succeeded together with the period of the three sacred originators at the dawn of our history. This country was governed by six emperors for 182 years. Despite such a short history, it was a great power dominating the continent.

The lineage of Korean sovereignty (국통맥) in BCE was succeeded by Hwanin Hwanguk, Hwanung Baedal, Dangun Joseon and Bukbuyeo, as presented in Table 4-1

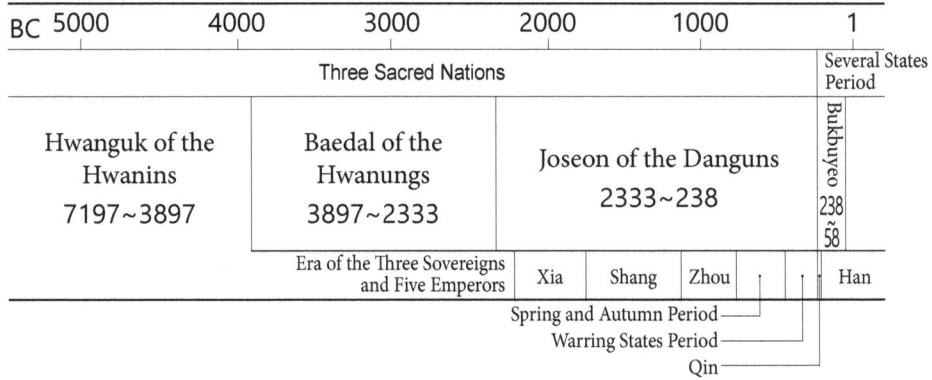

Table 4-1 Lineage of Korean Sovereignty in BCE

Hwanung Baedal is surely a part of Korean history, but Hwanin Hwanguk should be regarded as a history common to Eurasian countries. After leaving Hwanguk, Hwanung founded the country of Baedal people at Taebaek (태백) mountain, while Bango built that of the Huaxia people at Samui mountain. For this reason, China can argue that Hwanguk is a part of their history. *Hwandangogi* proves that Bango, worshiped as a god among Chinese people, was a real person.

As the weather was mild around Cheonsan mountain about 10,000 years ago, we can reasonably believe in the existence of Hwanin Hwanguk. Nevertheless, along with the great size of the country, I feel the 3301 years of history governed by seven Hwanins require some explanation. According to *Hwandangogi*, Hwanguk was divided into 12 smaller countries. For this reason, Hwanin Hwanguk may be regarded as a symbolic country.

The kings and tribal chiefs may have respected those Hwanins as their spiritual mainstays. If so, they thought themselves a part of Hwanguk. Moreover, if some of those seven Hwanins were actually families, then 3,301 years of reign are also a realistic possibility. Due to the scarcity of data, we should readily consider every possible scenario when it comes to ancient Korean history.

Korean history began with Gaecheon

Hwanung and his three subjects Pungbaek, Woosa and Woonsa led 3,000 heavenly descendants down to Taebaek mountain. This event is called Gaecheon or Heavenly Opening (개천). The Tiger people and the Bear people were native tribes, earthly descendants. Both wanted to be accepted as heavenly descendants. Hwanung said that they would be accepted only if they disciplined themselves by eating only mugwort and garlic. The Tiger people failed the test, while the Bear people succeeded. Then came the glorious end: the queen of the Bear people became the wife of Hwanung.

Koreans' belief in heavenly posterity can be felt in the act of the heavenly descendants (천손) civilizing the earthly descendants (지손). This is the foundational idea of Hwanung Baedal, that is, Hongik (홍익), meaning "humanitarian spirit." To the question "Who are we?", we should respond, "We are the descendants of heaven." To the question "How are we to live?", the correct answer is, "We should advance the welfare of mankind." Now you realize that the ideas of Gaecheon, Hongik, and heavenly posterity are all one and the same.

We cannot fully explain our country without the idea of heaven. Who are the heavenly descendants? They are those who try to follow heaven's will, in which case it is necessary to *know* heaven's will. Accordingly, all heavenly descendants should be eager to educate themselves. To summarize, the heavenly descendants are those determined to study and practice the providence of heaven. Some people misunderstand that the idea of heavenly posterity asserts the superiority of Koreans over other ethnicities. Rather, this idea is essentially the spirit pursued by the Korean people who study heaven.

Do you know why Gaecheonjeol (개천절), Korea's national foundation day, falls on the 3rd of October? This is because *Hwandangogi* reports that a Dangun named Wanggeom founded Joseon on the 3rd of October in the lunar calendar. The report can be summarized as, "A man called Wanggeom was crowed as Dangun on the 3rd of October in the 1565th year since Gaecheon."

We should pay attention to "the 1565th year since Gaecheon" in the above record. Dangun's foundation of Joseon is not the event called Gaecheon. Rather, the real Gaecheon occurred 1564 years before Dangun's foundation, when Hwangung built Baedal. This is an undeniably clear statement. That is, the 1565th year since Gaecheon is the first year of Dangi (단기) and corresponds to BCE 2333.

As we do not celebrate the real Gaecheon on the national foundation day, the entire 1565-year history of Hwanung Baedal is completely omitted from our history. Furthermore, another widespread misunderstanding is that the hero of the national foundation day is Dangun of Joseon, not Hwanung of Baedal. As a corollary, Dangun has become a familiar character to us, but we do not have any familiar image of Hwanung.

Taeho Bokhui and Chiu (치우) are the most remarkable heroes in the era of Hwangung Baedal. In particular, Bokhui is the originator of the cosmology of yin, yang and five elements and made Taegeuk. Thus, we have the world's oldest national flag dating back as many as 5,500 years ago. Bokhui went to the west and set the foundation of Chinese culture. To the Huaxia people, he was like a god from the east.

Well-known as the "red devil," the mascot of the South Korean national football team supporters, Chiu was Jaoji, the 14th Hwanung of Baedal. It is widely known that Chiu is called Cheonja, the son of heaven, in a footnote of Sima Qian's Records of the Grand Historian. According to records, every Chinese emperor worshiped Chiu. Before going to war, generals would hold a memorial rite for Chiu. He was the great hero of Koreans who moved the capital city of Baedal from Shinsi of Baekdu mountain to Cheonggu in order to conquer the western land. In Korea, the national football team support group the Red Devils preserves the image of Chiu, as illustrated in Figure 4-4.

Figure 4-4 Chiu, the mascot of the Red Devils

Sun-Ji Lee, an astronomer of Joseon, published a compilation of ancient records at King Sejong's order. The book, titled *Cheonmoonryucho* (천문류초), delivers the following record of the gathering of five major planets during the period of Hwanung Baedal

Ilwol Oseong Gaehap Jaeja (일월오성개합재자)

This means that the five major planets gathered with the moon and the sun. Chinese history includes the legendary era of Three Sovereigns and Five Emperors (삼황오제). The three sovereigns were Taeho Bokhui, Yeomje Sinnong (염제신농) and Hwangje Heonwon (황제헌훤), and the five emperors were Sohogeumcheon (소호금천), Jeonwookgoyang (전욱고양), Jegokgosin (제곡고신), Yao (요) and Shun (순). The record says that the planets gathered in 2467 BCE in the era of Jeonwookgoyang.

Using astronomical software, we can see that the five planets were arranged from left to right in the order of Mercury, Saturn, Venus, Jupiter and Mars in the eastern sky in the early morning of early September in 2470 BC, as shown in Figure 4-5.

Figure 4-5 Simulation of the alignment of the five planets
(created by Soo Jin Park)

Despite the difference of three years, it is reasonable for us to conclude that this is the alignment of the five planets, since we are estimating events dating back as many as about 4500 years. Thus, the record proves that the astronomical observatory during the period of Three Sovereigns and Five Emperors is not fiction but history. Moreover, as the astronomical records are found in the book published in Korea, this supports the assertion that the three sovereigns and five emperors, as well as the feudal lords of the Xia, Shang and Zhou dynasties, were all Baedal people. At the same time, it indirectly proves that the history of Hwanung Baedal belonging to the same period cannot be fiction.

The excavation of the Hongshan civilization's ruins was a historical event that captured international attention. Those ruins were first discovered in Liaoning province in the last century and date back 1,000 to 2,000 years earlier than the Yellow River civilization. As the mountains around the remains look red like the surface of Mars due to abundant iron ores, the civilization of the area is called Hongshan (red mountain, 홍산).

As shown in Figure 4-6, various jade ornaments and artifacts were excavated. Even jade silkworms were found as Figure 4-7. This implies that people had already been wearing silk clothes in the period of Hwanung Baedal. As the jade sword has the same shape as the lute-shaped bronze sword, the culture of Dangun Joseon must have originated from the Hongshan civilization. Many other relics prove that the Hongshan civilization belongs to the history of the Baedal people.

Figure 4-6 Tomb of Hongshan Civilization

Figure 4-7 Jade silkworm excavated from Hongshan

Emergence of Juyeok

In the history of China, King Yao of Tang (당요), King Shun of Yu (우순), King Yu of Xia (하우), King Tang of Shang (상탕), King Wen of Zhou (주문), King Wu of Zhou (주무) and his brother, the duke of Zhou (주공) are listed as saints. While the duke of Zhou was not a king, he is considered a saint because he made great contributions to complete the Juyeok (I Ching in Chinese, 주역). All of them, of course, are Baedal people. King Yao of Tang and King Shun of Yu are the last figures included in the Three Sovereigns and Five Emperors.

King Yu of Xia, King Tang of Shang, King Wen of Zhou, Duke of Zhou (From left)

The Xia Dynasty founded by King Yu survived for over 400 years. King Geol of Xia (걸왕), notorious for his debauchery, is one of the most famous tyrants and first representative "pond of alcohol and grove of meat" (주지육림). King Tang of Shang defeated King Geol and destroyed the Xia Dynasty with the help of Dangun Joseon. At that time, Dangun Joseon had the most outstanding figure called Yuwija, whose disciples including Yun Lee helped King Tang.

The Shang Dynasty continued for over 600 years but finally produced King Zhou (주왕), known as the second representative pond of alcohol and grove of meat. He killed his uncle Bigan (비간), who had advised him to be a good ruler, and confined King Wen of Zhou in a castle.

King Wen, who left the eight trigrams to us, invited Taegong Kang to serve as his tactician but died in vain without achieving his dream. However, his son King Wu, together with Taegong Kang, successfully carried out King Wen's will to defeat King Zhou and destroyed the Shang Dynasty. The duke of Zhou, a younger brother of King Wu, completed the basic scheme of the Juyeok by adding Hyosa (효사, interpretations of lines) to Goeisa (괘사, interpretations of trigrams), which had been left by King Wen.

As I mentioned above, the first emperor Qin began the history of the Huaxia people in the Central Plain of China. Thus, the Juyeok was actually made by the Baedal people and does not belong to Chinese civilization; the Baedal people are the rightful owner of ancient East Asian culture.

Sixty-four hexagrams constitute the basic scheme of the Juyeok. Pairs of the eight trigrams are arranged on top and bottom, resulting in a total of 64 hexagrams (= 8 x 8). As shown in Figure 4-8, the 64 hexagrams can be arranged in a square. King Wen annotated each hexagram; his annotations are called Goeisa (literally meaning annotation on trigram). Hence, there is a total of 64 annotations.

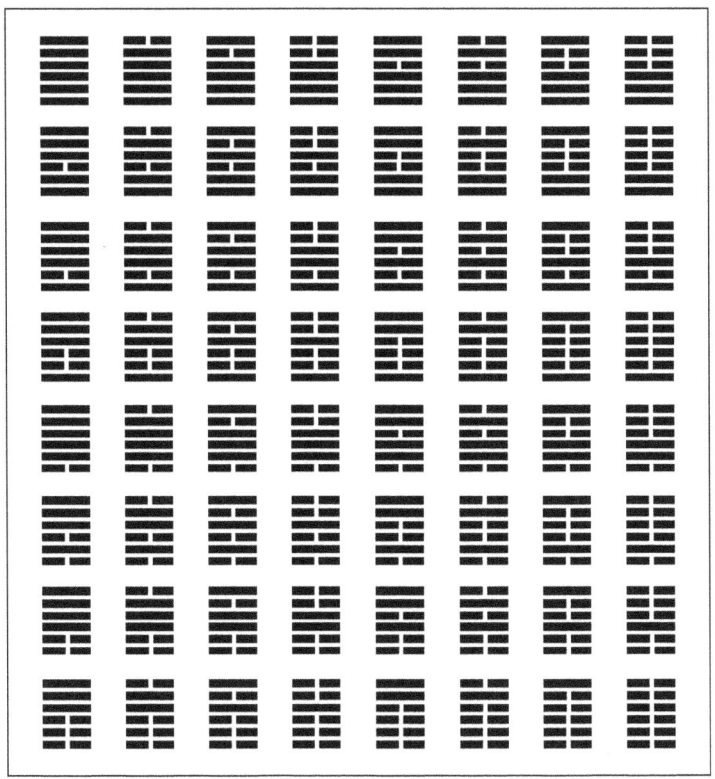

Figure 4-8 64 hexagrams of the Juyeok

In Figure 4-8, the lower trigram of the first eight hexagrams is Geon. In the following hexagrams, the lower trigrams are Tae, Li, Jin, Son, Gam, Gan and Gon, in that order. Once the lower trigrams are determined, the upper trigrams are arranged in the order of Geon, Tae, Li, Jin, Son, Gam, Gan and Gon. Thus, the 64 hexagrams are obtained. They can be arranged in such a beautiful circle as in Figure 4-9.

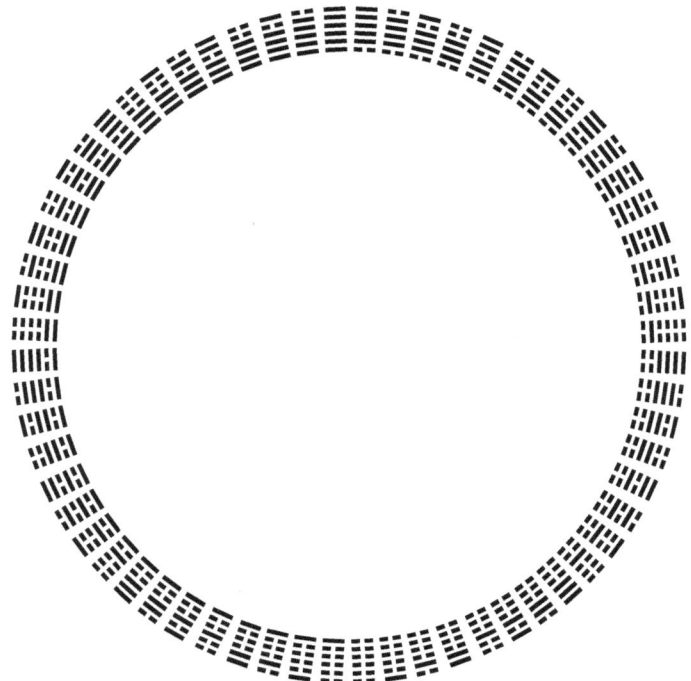

Figure 4-9 64 hexagrams in a circle

For example, consider the hexagram in which the upper trigram is Li and the lower one is Jin.

In the Goeisa, the interpretation of King Wen of this hexagram means "to be prosperous and beneficial by doing right," "to abstain from visiting a place where one is expected to be," "the sovereign should have help from his lords," and the like.

The Duke of Zhou, a son of King Wen, added annotations to the existing ones. He annotated Hyosa, each of six horizontal lines of each hexagram. Accordingly, there are a total of 384 annotations on hyo (= 64×6).

According to the Juyeok, any divination sign can be obtained merely by throwing a Yut stick six times. For a clear understanding of the cosmology of the Baedal people, Table 4-2 presents Cheonbugyeong, Hado, Taeho Bokhui's Eight Trigrams, Nakseo, King Wen's Eight Trigrams, and the Juyeok matched to the timeline of Table 4-1.

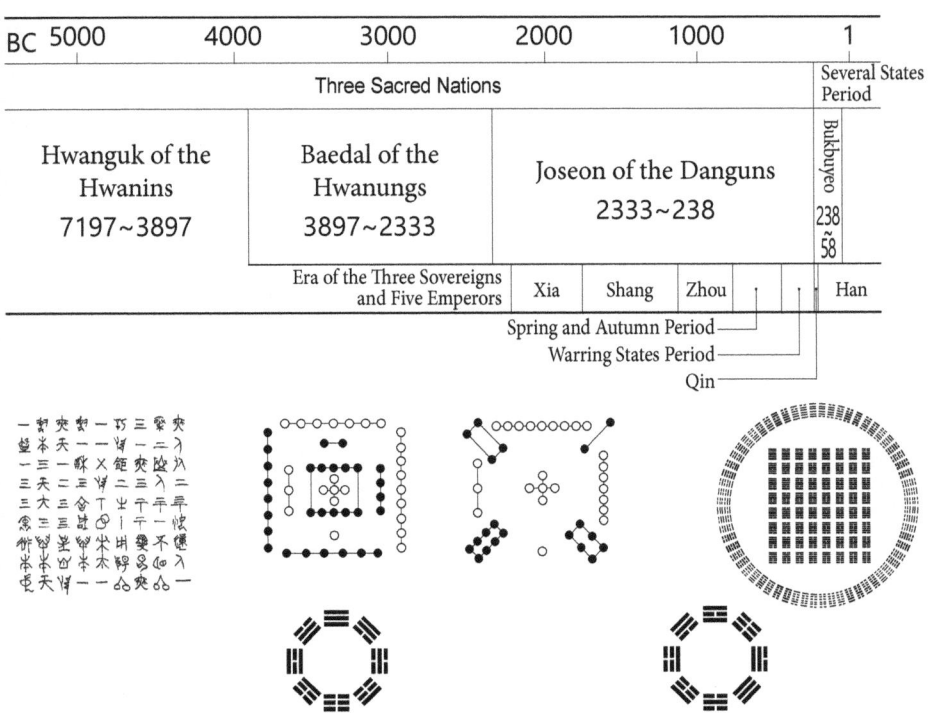

Table 4-2 Cosmology of the Korean people

Cosmic God and His Disciples: The Gods' Band

Cosmic god: "Who knew Korea had such a great history?"
Galaxy god: 'You're reading *Hwandangogi*? I've already read it…'

Cosmic god and Galaxy god:
    "When the heaven opened and Taegeuk danced…!"
Earth god: 'The Gaecheon Song is always such a blast!'

Chapter 5

Cheonsang yeolcha bunya jido, Korean Crowning Achievement

The Cheomseongdae observatory of Silla is one of the most marvelous astronomical achievements in Korean history. One cannot leave out the tomb murals of Goguryeo, which describe the three-legged crow representing the sun, and the rabbit and toad representing the moon and the 28 constellations. Baekjae introduced astronomy to Japan. These traditions of the Three Kingdoms Period were passed on to Seowoonkwan of Goryeo and Gwansanggam of Joseon, which at last produced Cheonsang yeolcha bunya jido, the cultural heritage of humanity.

Astronomy during the Three Kingdoms period and the Goryeo dynasty

The Cheomseongdae (첨성대) observatory of Silla (신라) is surely one of the most marvelous achievements in Korean history. The central window separates the observatory into 12 layers of stones above and below the aperture. There are around 360 stones, regardless of whether those in the window frame are included. Of course, our ancestors did not

Figure 5-1 Cheomseongdae observatory

simply heap up stones at random and coincidentally obtain a structure of 12 layers and 360 stones. Cheomseongdae is the very symbol of the universe, one that our ancestors built based on a conscious plan employing the codes of the universe. By developing an eye for heaven, we can realize our ancestors' insight.

We tend to call Cheomseongdae the oldest observatory in Asia. In fact, it may be the oldest one in the world. Cheomseongdae seems to be the most remarkable site among the astonishing relics in Gyeongju (경주). Though we see many wonderful temples and statues of Buddha in other countries, there are rarely observatories. Among the artifacts inherited from Silla dynasty are fragments of a sundial made of granite. *Samguksa* (삼국사), the Three Kingdoms Chronicle reports that a water clock was made and maintained by astronomers. There are many records of astronomical observations, including solar and lunar eclipses, comets and shooting stars.

Figure 5-2 Fragment of sundial of Silla dynasty

The culture of heavenly posterity reached its peak during the Goguryeo dynasty. Many images and writings on the tomb murals display the astronomy of Goguryeo. The blue dragon, white tiger, red phoenix and black turtle, which lay the foundation of the 28 constellations, commonly occupy the east, west, south and north walls, respectively. In particular, the tomb located at Deokheung-ri, Gangseo-gun, Pyeongannam-do is a perfect time capsule.

The Big Dipper and South Dipper (남두육성) on the north and south walls, respectively, are the reference constellations of the sky. As the sun with the three-legged crow is on the east wall and the moon with the toad is on the west wall, they remind us of the image of the sun, moon and five mountains. For example, Figure 5-3 shows the mural on the south wall. The South Dipper is on the upper-right of the wall. The galaxy is in the center between Gyeonwoo and Jingnyeo. There is also an animal with a human face right below Jingnyeo.

Figure 5-3 The south mural of the tomb of Deokheung-ri

Gwanreuk (관륵) of Baekjae (백제) introduced astronomy to Japan. The Japanese people agree with this historical fact, which provides indisputable evidence that Baekjae developed its own astronomy. Unified Silla succeeded the astronomical tradition of the Three Kingdoms Period. Navigation cannot be developed without the development of astronomy. Bo-Go Jang of Unified Silla could not have navigated to India after passing through China. In this regard, we cannot even imagine the greatness of Korea's lost astronomy and maritime history.

The cosmic view established long before the Three Kingdoms Period imagined that the three-legged crow was in the sun and the rabbit or the toad lived on the moon. The Goguryeo artifact in Figure 5-4 is the three-legged crow. This three-legged crow was not considered a normal crow but was worshipped as the sun god living in the sun. In short, astronomy was so developed during the Three Kingdoms Period that *Samguksa*: Three Kingdoms Chronicle includes as many as 240 records about such phenomena as solar and lunar eclipses, comets and shooting stars.

Figure 5-4 Three-legged crow of Goguryeo artifact

Figure 5-5 Bronze plaques of the three-legged crow (left) and the rabbit and toad (right) symbolizing the sun and the moon, respectively

Cheomseongdae of Silla was succeeded by Seowoonkwan (서운관) of Goryeo. The remains of Seowoonkwan observatory are preserved near Gaeseong. Goryeo obtained its status as a maritime kingdom in virtue of its excellent achievements in astronomy. The astronomy volume *Goryeosa* contains about 5,000 astronomical records. The mural of 28 constellations in the tomb of Seosam-dong, Andong-si and that of the Big Dipper in the tomb of Seogok-ri, Paju-si are particularly famous among the astronomical remains of the Goryeo dynasty,.

Figure 5-6 Remains of an observatory built in Goryeo Dynasty

Cheonsang yeolcha bunya jido

Taejo Seong-gye Lee, who destroyed Goryeo and founded a new dynasty, Joseon, in 1392, wanted the people to accept the new dynasty as heaven's will. After acquiring a rubbed copy of the star map of Goguryeo, he was so pleased that he ordered it to be inscribed on stone. In 1395, the fourth year of King Taejo, the inscription was completed. This is Cheonsang yeolcha bunya jido (천상열차분야지도), national treasure No. 228, now preserved in Gyeongbok Palace. Cheonsang yeolcha bunya jido means "a figure representing the faces of heaven by order and by section."

In Cheonsang yeolcha bunya jido, as shown in Figure 5-7, a total of 1467 stars is inscribed with different sizes according to brightness. Cheonsang yeolcha bunya jido consists of three enclosures and 28 constellations.

There is a figure called Honhyojungseongdo (혼효중성도) directly above the star chart. This figure lists the constellations rising in the sky between the evening and early morning according to the 24 seasonal divisions. For example, the winter solstice in the center at the bottom of Figure 5-8 has the following inscription:

hon sil hyo jin jung 혼실효진중

This means that the winter solstice is when Sil (Pegasus α) is in the southern sky in the evening and Jin (Corvus γ) rises in the southern sky in the early morning. Thus, Honhyojungseongdo tells when each of the 24 seasonal divisions comes.

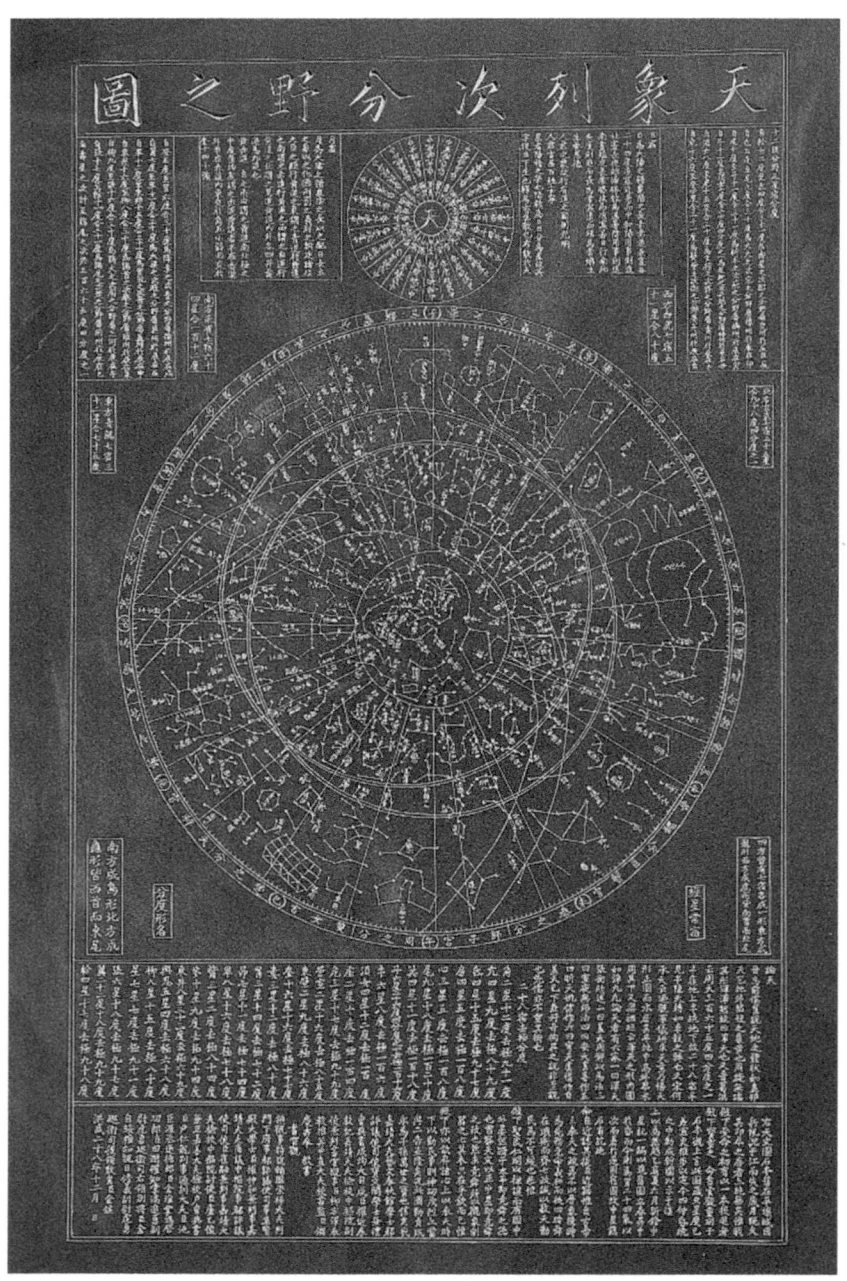

Figure 5-7 Cheonsang yeolcha bunya jido

Figure 5-8 Honhyojungseongdo

The ball-shaped sky over our heads is called the celestial sphere. Figure 5-9(A) describes the concept of the celestial sphere, including the Earth's orbit. As the orbit is tilted 23.5 degrees from the plane of the solar system, the ecliptic and the equator also cross each other at 23.5 degrees.

If you take a close look at Figure 5-9(A), you will find two circles. The red one is the equator and the yellow one is the ecliptic. The celestial heaven of Figure 5-9(A) is projected onto Figure 5-9(B), with the north pole at the center. You may think that the bottom of the celestial heaven of Figure 5-9(A) has been ironed out. But as shown in Figure 5-9(B), the equator becomes a concentric circle, while the ecliptic does not. Figure 5-9(B) explains the two circles in the center of Figure 5-7.

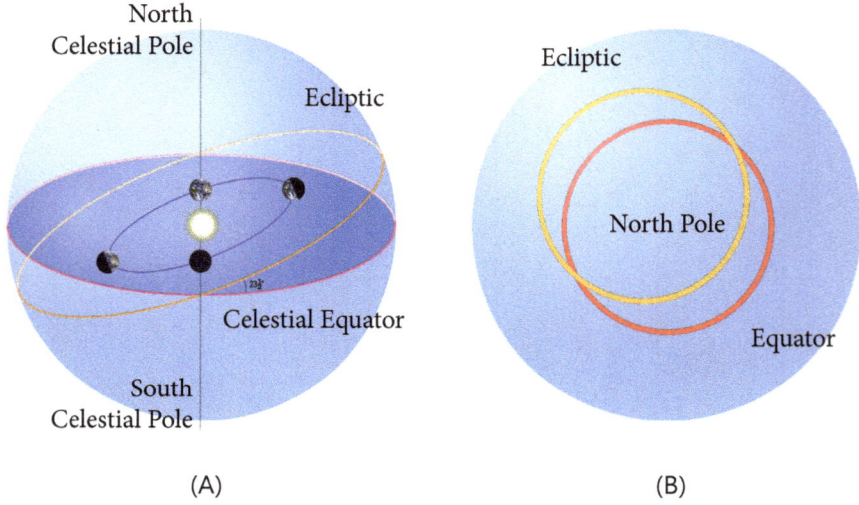

Figure 5-9 Basic scheme of Cheonsang yeolcha bunya jido

The inscription at the bottom right of the star chart states that the original work had been preserved in Pyongyang Castle but was lost in the river during a war. The original star chart is a cultural heritage as precious as the Monument of King Gwanggaeto the Great (광개토태왕). This is because the monument demonstrates the power of the Korean people, while the original work shows the cultural level attained by Koreans. The latter is one of the most precious natural treasures since it proves that we are the posterity of heaven. We cannot be sure whether the Pyongyang Castle noted in the inscription refers to present-day Pyongyang. We may find the answer after examining every corner of the Daedong River someday.

An inscription in Cheonsang yeolcha bunya jido reports that Geun Gwon, one of the founders of the Joseon dynasty, composed the piece, Bang-Taek Ryu took on the task of astronomical calculation, and Gyeong-Soo Seol wrote the piece. Composing and writing inscriptions could be done by many scholars, but only a few could correct the constellations, which had been changed since the rubbed copy of Goguryeo Dynasty. In other words, Cheonsang yeolcha bunya jido could not have been completed without Bang-Taek Ryu.

Portrait of Bang-Taek Ryu (provided by Yong-Jin Cho)

King Sejong the Great

King Sejong the Great of Joseon Dynasty was pained by the reality that the astronomers of China were not able to accurately predict the astronomical phenomena appearing in the sky of Joseon. As Chinese astronomy was based on the observations of the Beijing sky, it was not accurately fit for that of Seoul. We can imagine King Sejong's agony as follows: "I am the King of this whole kingdom. However, I cannot predict what happens in the sky of this country. How disappointing it is!" His agony was that of all Korean people.

King Sejong the Great

One day, King Sejong the Great waited in formal attire under the burning sun for a solar eclipse. The event of the moon blocking the sun, which is the symbol of the King, had many implications in the days of the Kingdom. Unfortunately, the solar eclipse occurred about 15 minutes later than expected. The king sentenced the official of astronomy to flogging. It is not difficult for us to imagine that King Sejong must have been more distressed than the official who was flogged.

King Sejong the Great ordered Sun-ji Lee and other astronomers to publish Chiljeongsan (칠정산), or Calculation of the Motions of the Seven Celestial Determinants, since he was dissatisfied with the requirement that Joseon had to send an envoy to China every winter solstice in order to receive the astronomical data for the coming year. Our own calendar system was established at last. I think this is one of King Sejong's greatest achievements, along with the creation of Korean characters, Hangul (한글).

Ganuidae (간의대), an astronomical platform of King Sejong's period, was restored to its original size and sits in front of the main building of the Korea Astronomy and Space Science Institute in Daejeon as shown in Figure 5-10. The original Ganuidae used to be installed at the northwest corner of Gyeongbokgung Palace.

Figure 5-10 Ganuidae, an astronomical instrument of King Sejong's period

During the Joseon Dynasty, Seowoonkwan was changed to Gwansanggam (관상감). Figure 5-11 shows the locations of Ganuidae and Gwansanggam in Gyeongbokgung Palace. Gwansanggam has a very important role since it was where the king made a direct appeal to heaven. Throughout the Joseon Dynasty, Younguijeong (영의정), the prime minister, had the additional title of the head of Seowoonkwan or Gwansanggam as many as nine times. This suggests much about the heavenly tradition of Koreans. In the late stage of Joseon Dynasty, Dae-yong Hong and other scholars left some astronomical findings. Among the remains, there is Gwancheondae (관천대) of Chnaggyeonggung Palace in Figure 5-12.

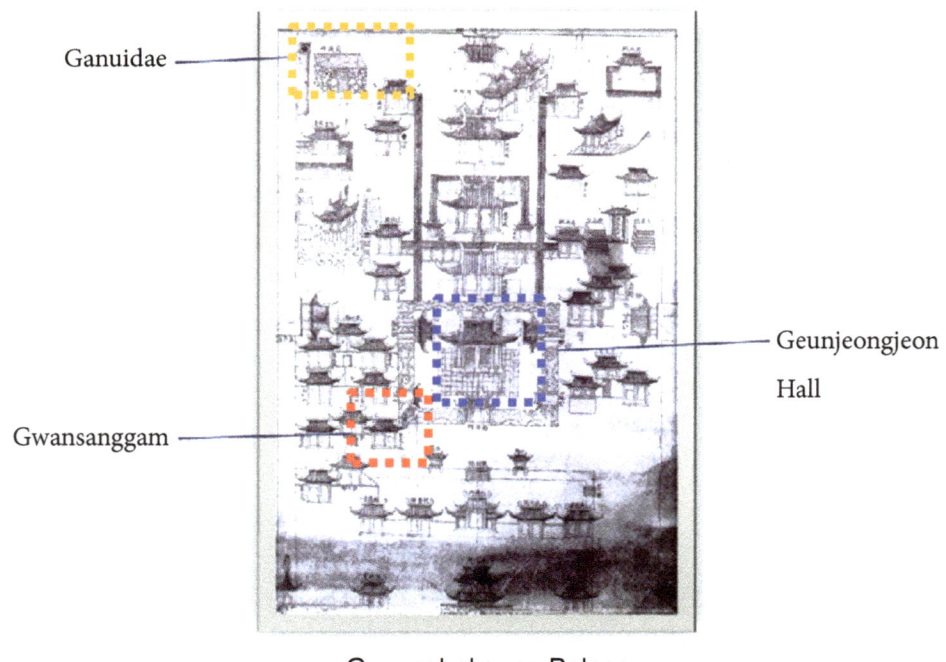

Figure 5-11 Locations of Gwansanggam and Ganuidae in Gyeongbokgung Palace

Figure 5-12 Gwancheondae of Changgyeonggung Palace

Cosmic God and His Disciples: Making a Supernova

Galaxy god: "It's time to make a supernova."
Cosmic god: "All right, is everything ready?"

Galaxy god: "Fire!"
Earth god: "Bullseye! Right on target!"

Chapter 6

The Astronomical Renaissance of Modern Europe

Until the Middle Ages, Eastern astronomy had been more advanced than that of the West. However, the emergence of astronomical telescopes in Europe made the East fall behind. The theory of geocentrism, doubted by Copernicus, was finally overturned by Galilei's astronomical observations using a telescope. Later, Newton discovered the force of gravity after studying Kepler's laws of planetary motion.

Heliocentric view of the universe

As I mentioned above, the Ancient Greeks not only knew that the earth was round but also calculated the earth's size relatively accurately. However, as astronomy regressed in the Middle Ages, people came to believe that the earth was flat. They thought that if they traveled too far, they would fall off a cliff at the end of the sea. Columbus confessed that he had to struggle against this idea before setting off to discover America. Magellan could not have circumnavigated the globe without believing that traveling in one direction would take him back to where he departed.

In Western cosmology during the Middle Ages, which is also called the Dark Ages, there was simply heaven and hell. In the Aristotelian universe, the sublunary realm at the center of the universe was the world of human beings and sin, while the superlunary realm was the world of God. Accordingly, any arguments against heliocentrism were labeled blasphemy. Meanwhile, as Western astronomy stagnated, astronomers in the Middle East observed solar and lunar eclipses and made star catalogues. They added many stars beginning with "Al" and others like the famous Betelgeuse in the constellation Orion.

Facing incessant challenges, geocentrism finally collapsed in the modern era and was replaced by heliocentrism as proposed by Copernicus and his advocates. As was shown in Figure 2-2, geocentrism explained the retrograde motions of planets by assuming that they were moving in smaller circles. On the other hand, heliocentrism does not require such movements.

Copernicus

Galileo observed Venus using a small telescope. As with the moon, we can see only the side of Venus that faces the sun. Figure 6-1 shows an image of Venus captured by a small telescope during the evening, when it is farthest from the sun and sets at a 48-degree angle. Venus appears similar to a half moon. As Venus looked like a half moon or had a more convex shape, Galilei was certain that geocentrism was wrong. This was because in the geocentric universe, only a crescent-shaped Venus was possible (Figure 2-2).

Galilei

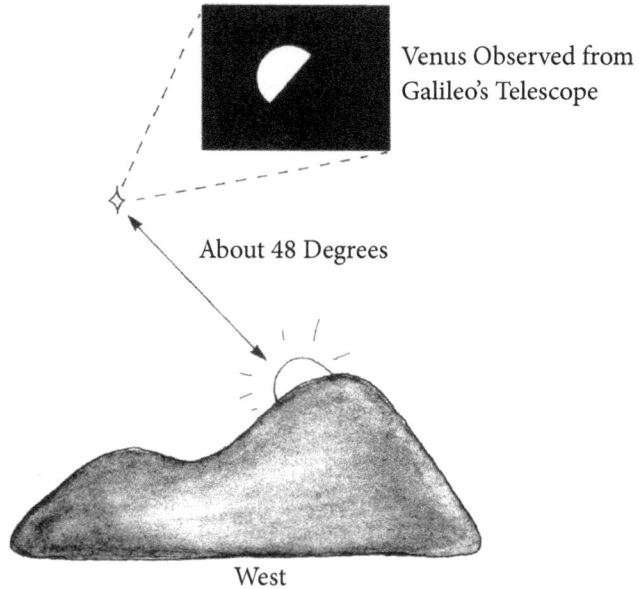

Figure 6-1 Half moon-shaped Venus

In addition, Galileo discovered the four bright moons of Jupiter as shown in Figure 6-2, in which Jupiter is in the center. The moons can all be on the left or right of Jupiter. Though each moon is as large as Mercury, they are reduced to the satellites of the solar system's largest planet. The moons are named after Io, Europa, Ganymede and Callisto, who were all loved by Zeus. As Jupiter is the Roman name of Zeus, they are bounded to Zeus forever.

Figure 6-2 Sketch of Jupiter and its four bright moons

Galileo's telescope shown in Figure 6-3 is not as good as those used in school laboratories. Nevertheless, he overturned geocentrism using this humble tool. From an educational perspective, Galileo's observations are very meaningful. Due to his beliefs, he was called before the Holy Office in a religious trial. Although he had to lie to save his life, he uttered the words "And yet it moves" as he stepped out of the court. What he meant is that Earth does, in fact, revolve around the sun.

Figure 6-3 Galileo's telescope

Discovery of gravity

In the heliocentric universe as we know it today, the sun is at the center of the solar system, and the planets revolve around the sun in the order of distance from the sun: Mercury, Venus, Earth, Mars, Jupiter and Saturn. The distance from the sun to Earth is about 150 million kilometers. This is the astronomical unit (AU). The distances from the sun to Mercury, Venus and Mars are 0.4AU, 0.7AU and 1.5AU, respectively. Jupiter is as far as 5.2AU from the sun, and Saturn is nearly twice that at 9.6AU. Hence, the closer to the sun, the closer the planets are to each other, and vice versa.

English	Korean	Sign	Orbit radius (AU)	Orbital period (year)
Mercury	수성	☿	0.4	0.24
Venus	금성	♀	0.7	0.62
Earth	지구	⊕	1	1
Mars	화성	♂	1.5	2
Jupiter	목성	♃	5.2	12
Saturn	토성	♄	9.6	30

Table 6-1 Five planets' orbits of revolution

The sun pulls in closer planets with a stronger gravitational force. Thus, the closer to the sun, the shorter the planet's orbital period. In this way, the planet overcomes the sun's gravity by increasing its own centrifugal force. In fact, the orbital periods of Mercury and Venus are as short as 88 days and 225 days, whereas those of Jupiter and Saturn are as long as 12 years and 30 years.

After studying planets, Kepler presented the following three laws of planetary motion:

① The Law of Ellipses: All planets move in elliptical orbits, with the sun at one focus.

② The Law of Equal Areas: A planet moves fastest when it is closest to the sun and slowest when it is furthest from the sun. Thus, a line drawn from a planet to the sun sweeps out equal areas in equal times.

③ The Law of Harmonies : If the semimajor axis of a planetary orbit is a and the orbital period is p, then every planet has the same p^2/a^3 ratio. That is, p^2/a^3 = constant

Kepler

The most astonishing of these is ①, the Law of Ellipses. From time immemorial up to the modern era, humans did not doubt the dogmatic belief that every heavenly body moved in circular orbits. Accordingly, this law was revolutionary. Nevertheless, as the planets' orbits are *nearly* circular despite actually being elliptical, this incorrect belief is not problematic as long as total accuracy is not required.

Unlike a circular orbit, an elliptical orbit continuously changes the distance between the planet and the sun. The sun pulls the planet with stronger gravitational force when it is closer and with weaker gravitational force when it is farther away. The planet revolves faster as it moves closer to the sun and more slowly as it moves farther away, thereby maintaining its orbit. This produces ②, the Law of Equal Areas. In Figure 6-4, it takes the same time for the planet to move from 1 to 2 and from 3 to 4, and the two dashed sectors have equal areas.

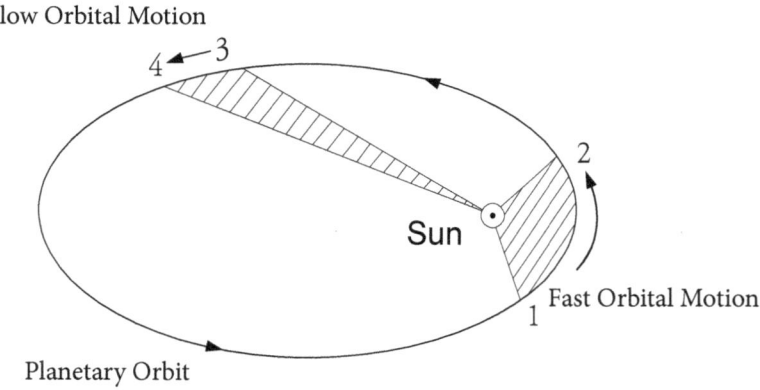

Figure 6-4 The Law of Equal Areas

If you consider the values given in Table 6-1, you may understand ③ the Law of Harmonies. For example, in the case of Mercury, a= 0.4AU, and p = 0.24 year, and p² = 0.0576, a³ = 0.064 are almost equal. That is,
$$\frac{p^2}{a^3} = 1$$

Table 6-2 applies this ratio to the other planets. If we apply accurate values, the values of the rightmost column should be exactly 1. Marveling at his finding, Kepler named it the Law of Harmonies.

Planet	P	P²	a	a³	P²/a³
Mercury	0.24	0.0576	0.4	0.064	0.9
Venus	0.62	0.3844	0.7	0.343	1.1207
Earth	1	1	1	1	1
Mars	2	4	1.5	3.375	1.1852
Jupiter	12	144	5.2	140.608	1.0241
Saturn	30	900	9.6	884.736	1.0172

Table 6-2 Kepler's Law of Harmonies

Sir Isaac Newton found that any two objects with mass attract each other, and that the magnitude of this attracting force is inversely proportional to the square of the distance. Otherwise, as he declared, Kepler's laws of planetary motion would not be true. If two objects placed at a distance have masses and, respectively, the force of gravity is given as follows:

$$F = G \frac{m_1 m_2}{r^2}$$

Here G represents the universal gravitation constant. Thus, the mighty law of universal gravitation was discovered. This theory of gravity was motivated by the insight that apples fall from trees, and the moon revolves around Earth on the same principle.

Newton tried to explain the universe filled with countless twinkling stars by using his theory of gravity. However, he failed because all the stars attracting each other must inevitably condense to the center of mass. Accordingly, Newton argued that the universe is infinite. If the universe has an infinite number of stars, there is no reason for any one star to be attracted in a specific direction. Thus, the universe can maintain its stability.

Newton

Completion of lunisolar calendar

As I introduced above, the Roman Emperor Julius Caesar ordered a new calendar consisting of 365 days in a year. In astronomy, since the exact duration of one solar year is 365.2422 days, almost one day is added every four years (0.2422 × 4 = 0.9688 days). To compensate for this, the Julius calendar introduced a leap year, with 366 days, occurring every fourth year. With leap years, however, 0.0312 days (= 1 - 0.9688) are added every four years. In other words, about three days are added over 400 years (0.0312 × 100 = 3.12).

To solve this problem, Pope Gregory XIII commissioned a new calendar in 1582 as shown in Table 6-3. This Gregorian calendar has a leap year every four years but skips leap years three times in 400 years. In other words, if the year is divisible by 4, that year becomes a leap year with 29 days in February. If the year is divisible by 100 again, then it becomes an ordinary year. If divisible by 400 again, it becomes a leap year.

Since 1800 and 1900 CE are divisible by both 4 and 100, they are ordinary years with 28 days in February. Since 2000 CE is divisible by 4, 100 and 400, it is a leap year with 29 days in February. The Gregorian calendar has an error of 0.0012 days every four years, which corresponds to about a day every 1,000 years. This is the solar calendar we use today.

January 1 of the current solar calendar has no astronomical significance. This day has nothing to do with the sun and the moon; it is simply a day in the middle of winter. On the other hand, the lunar calendar seems to reflect the season to a certain extent since Ipchun falls in January. Therefore, according to the lunar calendar, spring begins in January.

Julian			Gregorian			Modern		
Month	Name	days	Month	Name	days	Month	Name	days
1	Janualis	31	1	Janualis	31	1	January	31
2	Februalis	29	2	Februalis	28	2	February	28
3	Martius	31	3	Martius	31	3	March	31
4	Aprilis	30	4	Aprilis	30	4	April	30
5	Maius	31	5	Maius	31	5	May	31
6	Junius	30	6	Junius	30	6	June	30
7	Julius	31	7	Julius	31	7	July	31
8	Sextilis	30	8	Augustus	31	8	August	31
9	September	31	9	September	30	9	September	30
10	October	30	10	October	31	10	October	31
11	November	31	11	November	30	11	November	30
12	December	30	12	December	31	12	December	31
		365			365			365

Table 6-3 Julian calendar and Gregorian calendar

The dates in the solar calendar for the 24 seasonal divisions hardly change. This is because the 24 seasonal divisions were established based on the solar calendar. As you see in Table 6-4, Ipchun and Usu fall in February by the solar calendar and January by the lunar calendar, and Gyeongchip and Chunbun are in March by the solar calendar and in February by the lunar calendar. In this way, each month has two of the 24 seasonal divisions. This is natural enough because one year comprises 12 months.

Seasonal divisions	Mid-month divisions	Solar calendar (immutable)	Lunar calendar (mutable)
Ipchun	Usu	February	January
Gyeongchip	Chunbun	March	February
Cheongmyeong	Gogu	April	March
Ipha	Soman	May	April
Mangjong	Haji	June	May
Soseo	Daeseo	July	June
Ipchu	Cheoseo	August	July
Baengno	Chubun	September	August
Hallo	Sanggang	October	September
Ipdong	Soseol	November	October
Daeseol	Dongji	December	November
Sohan	Daehan	January	December

Table 6-4 24 seasonal divisions

However, when using the lunar calendar, two seasonal divisions are not always allocated to each month. This is because a leap month may push one further back. On the other hand, the solar calendar always assigns two seasonal divisions to each month. For example, Gyeongchip and Chunbun always fall in March. Chunbun always falls around March 21 but never on March 25 or a similar day. Thus, you may understand that the 24 seasonal divisions are based on the solar calendar.

Until the Joseon period, our ancestors did not know about the Julian or Gregorian calendars but employed the lunar calendar, including the 24 seasonal divisions. This is called the lunisolar calendar. For instance, November in the lunar calendar is called the month of Dongji because Dongji never fails to fall in that month. Actually, the 24 seasonal divisions are divided into 12 seasonal divisions and 12 mid-month divisions. A more proper name would be the 24 seasonal and mid-month divisions.

As the lunar and solar calendars are used together, leap months occur which have no mid-month divisions. Any month having only one seasonal division and no mid-month division becomes a leap month. This method is called Mujungchiyunbeop (무중치윤법), which literally means a method of handling months with no mid-month division by using a leap month.

Cosmic God and His Disciples: Making Planets

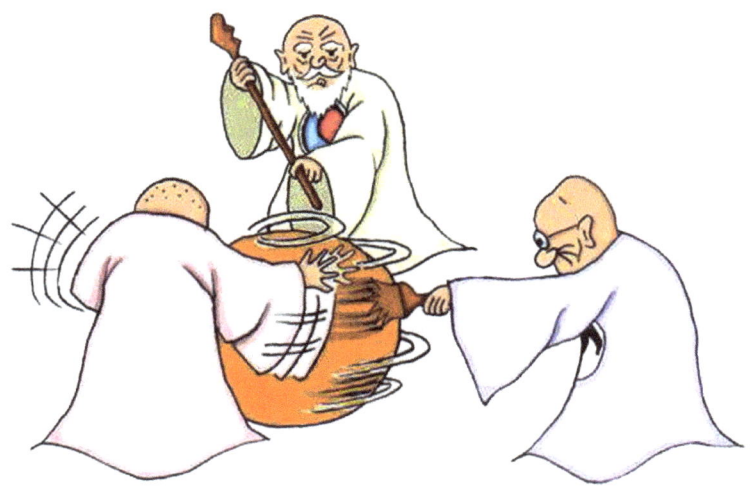

Cosmic god: 'The red spot needs to be just right…'
Galaxy god: 'The stripes should look natural.'
Earth god: 'My arm hurts. I've been spinning this for ages.'

Cosmic god: "Hi-yah!"
Galaxy god: 'He's making the rings of Saturn.'
Earth god: 'He only managed to get one…'

Chapter 7

Great Universe Unveiled

Herschel discovered Uranus using a large astronomical telescope. He also observed that the galaxy is composed of countless stars. Shapley was the first to argue that our solar system is at the edge of a galaxy called the Milky Way. In the East, Il-bu Kim set right the Juyeok with his Ideology of Jeongyeok. Only two centuries ago, human beings came to realize that the Milky Way is just one among an infinite number of galaxies in a vast universe.

Uranus, Neptune and Pluto

In the modern period of the West, Herschel made an astronomical telescope of 50cm in diameter to observe space. He discovered the sixth planet, which was the most shocking discovery in the history of astronomy to that date. The dogma of the five planets, which had been deep-rooted both in the East and in the West for thousands of years, collapsed at last.

Herschel

Figure 7-1 Herschel's telescope

Soon, the sixth planet was named after Uranus, the god of heaven in Greek mythology. The names of the planets Neptune and Pluto came from the Greek gods of the sea and the underworld, respectively.

Later, Herschel also discovered that the Milky Way was a group of countless stars, expanding the horizon of human imagination beyond the solar system to the galaxy. Thus, Western astronomy began to progress beyond Eastern astronomy. We can guess a similar situation occurred in medicine and biology after the microscope was invented in the West.

Herschel was not able to observe Neptune, as it is darker than Uranus. However, after observing the movement of Uranus, many astronomers were certain that there would be another planet beyond it. The winner of this race to discover the new planet was Galle.

Pluto was first discovered in 1930 by Tombaugh, who was Lowell's assistant. Lowell visited Joseon in the 19th century. He was impressed by thatched-roof houses dimly visible through the fog and described Joseon as "the Land of Morning Calm." Many people mistakenly attribute this phrase to the Indian poet Tagore, who eulogized Joseon as "the Lamp of the East".

Later, Pluto was demoted from its status of planet in the solar system. From the beginning, Pluto's status was in doubt. The four innermost planets of the solar system — Mercury, Venus, Earth and Mars — are relatively small and their surfaces are covered with soil. The outer planets of Jupiter, Saturn, Uranus and Neptune are larger and consist mostly of fluids. However, Pluto seems to be exceptional; it is smaller than Mercury and its orbit of revolution is tilted 17 degrees from the plane of the solar system. Moreover, Pluto's elliptical orbit is so eccentric that it is sometimes closer to the sun than Neptune.

Dispmayed by these aberrant qualities, some astronomers tried to strip Pluto of the status of planet at every opportunity. When it was discovered in 1978 that Pluto had a satellite, such attempts seemed to be frustrated. This satellite, named Charon after the ferryman of the dead in Greek mythology, saved Pluto. Charon perfectly fit as its satellite.

However, as more and more small heavenly bodies like Pluto were discovered due to the advancement of astronomical telescopes, the tragedy began. As long as Pluto was counted as one of the planets, many other similar newly discovered bodies might also need to be given the status of planet. The International Astronomical Union took the easy way out, demoting Pluto. It was the ideal solution to all of these problems.

Table 7-1 presents the basic features of the planets in the solar system. Based on this table, their orbits of revolution can be described as in Figure 7-2. You may realize that Mercury, Venus, Earth and Mars revolve around the sun at a close distance to each other, while Jupiter, Saturn, Uranus and Neptune are far more distant from each other.

English	Korean	Sign	Orbit radius (AU)	Orbital period (year)
Mercury	수성	☿	0.4	0.24
Venus	금성	♀	0.7	0.62
Earth	지구	⊕	1	1
Mars	화성	♂	1.5	2
Jupiter	목성	♃	5.2	12
Saturn	토성	♄	9.6	30
Uranus	천왕성	♅	19.2	84
Neptune	해왕성	♆	30	165

Table 7-1 Features of the revolutional orbits of the planets

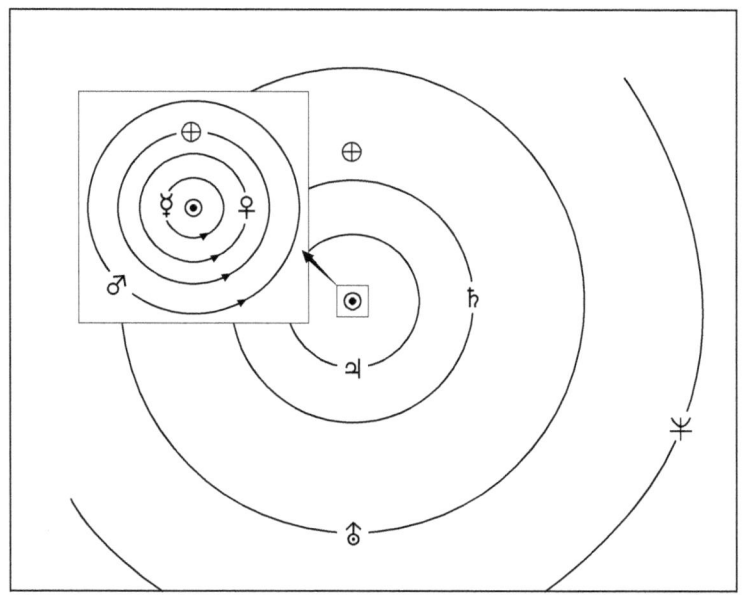

Figure 7-2 Solar system

From Juyeok to Jeongyeok

It is often said that the Juyeok is also called Zhouyi (Changes of Zhou), and thus has Chinese origin. However, as we saw above, the ruling classes of the Xia, Shang and Zhou dynasties were Baedal people. For this reason, the Juyeok is our own tradition, and both the King Wen and the Duke of Zhou were of Baedal origin. Authors who mistakenly attribute the Juyeok to China unconsciously imbue their books with Sino-centrism. The ruling class of Joseon, who were well-versed in the Juyeok, were no exception. Even they accepted Sino-centrism as the vision of the country. Let me stress this again: Eastern astronomy is nothing but the astronomy of Baedal people. I think that the existing books on the Juyeok need to be fundamentally revised by adding the history of Hwandangogi.

Also, as some authors on the Juyeok assert that Taeho Bokhui's eight trigrams correspond to Early Heaven and the King Wen's eight trigrams to Later Heaven, they have weakened the meaning of the Korean people's thought on Later Heaven Gaebyeok (개벽). They argue that Taeho Bokhui's eight trigrams and Hado describe the early heaven universe and that King Wen's eight trigrams and Nakseo describe the later heaven universe. However, as I pointed out above, the latter pair—King Wen's trigrams and Nakseo—have a clear relationship, but the former pair does not. Those authors' misunderstanding is attributable to the fact that Taeho Bokhui made both his eight trigrams and Hado.

I said before that Taeho Bokhui's eight trigrams are spatial arrangements and King Wen's eight trigrams are temporal ones. Just as the theory of relativity presented four-dimensional space-time by combining three-dimensional space and one-dimensional time, Il-Bu Kim in the late Joseon period proposed a new set of eight trigrams by combining those of Taeho Bokhui and King Wen. Unlike the Juyeok, Il-Bu Kim's new Juyeok includes many scientific interpretations, since it was written in the 19th century when the modern scientific achievements of the West had been introduced.

Il-Bu Kim

Il-Bu Kim's Juyeok is called Jeongyeok (정역), and the eight trigrams derived from it are the eight trigrams of Jeongyeok. Some authors on the Juyeok are so Sino-centric that they ignore Jeongyeok. They either do not acknowledge it or are ignorant of it.

We should understand that Taeho Bokhui and King Wen's respective eight trigrams represent the early heaven, while the eight trigrams of Jeongyeok represent the later heaven.

As shown in Figure 7-3, Jeongyeok's eight trigrams are 1 Son, 3 Tae, 4 Gam, 5 Gon, 6 Jin, 8 Gan, 9 Li and 10 Geon. These trigrams have 9 and 10 instead of 2 and 7. In Jeongyeok, 2 and 7 have separate forms of 2 Cheon and 7 Ji. Like in Taeho Bokhui's eight trigrams, the total number of lines in any two trigrams facing each other is 9, and the trigrams of Geon and Gon take the center position. However, "heaven and earth (천지)" is replaced by "earth and heaven (지천)".

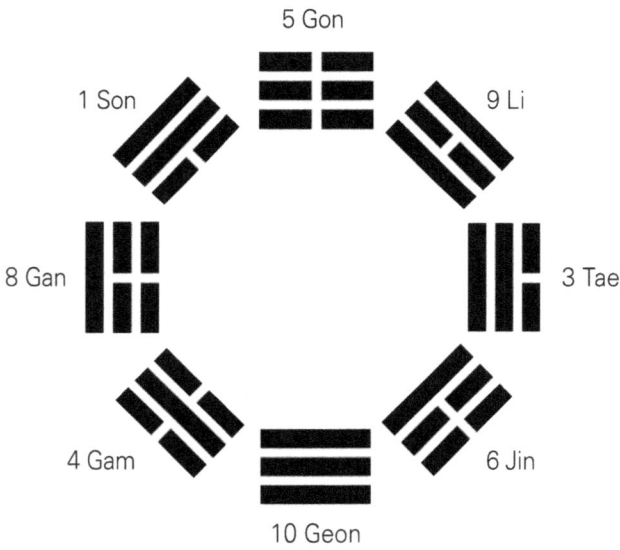

Figure 7-3 Eight trigrams of Jeongyeok

Jeongyeok began "Season 2" of Eastern astronomy. In other words, it marks the expansion from Table 4-2 to Table 7-2. Jeongyeok has a peculiar view of universe based on three principles: Mugeuk (무극), Taegeuk (태극) and Hwangeuk (황극). As I mentioned before, since Jeongyeok reflects abundant astronomical interpretations including the lunisolar calendar, it is still being studied actively.

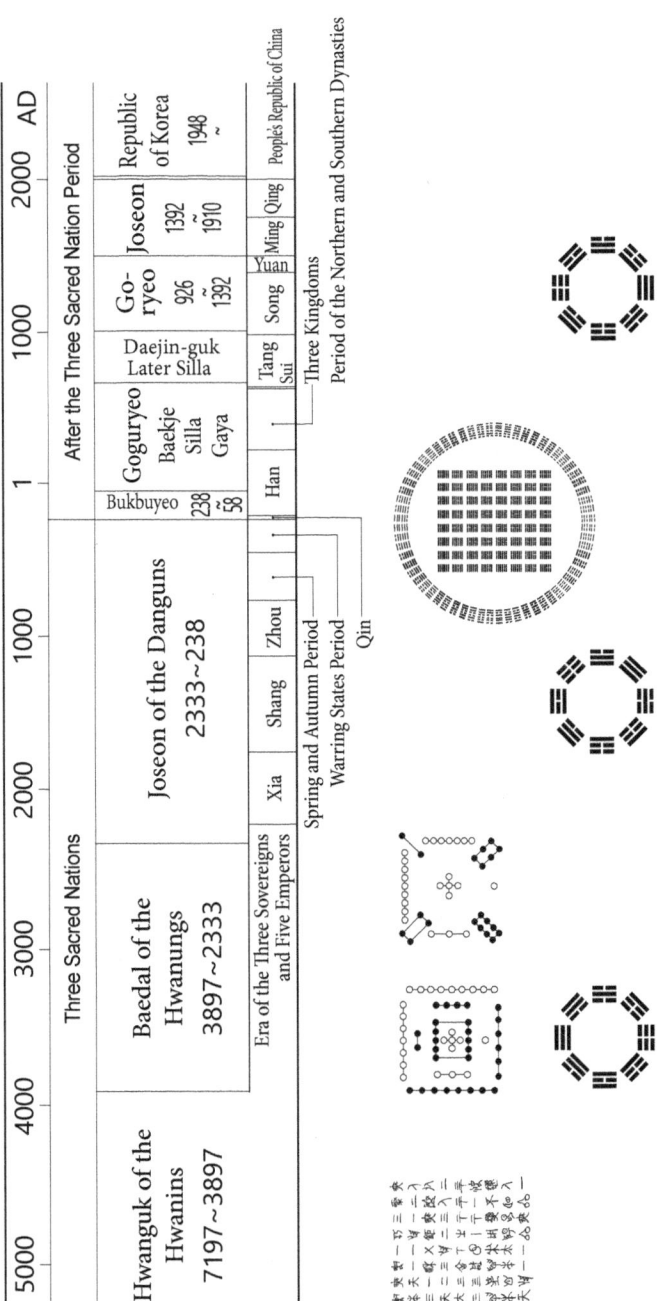

Table 7-2 Cosmology of the Korean people

The Donghak (동학) Revolution has been disparaged as a mere rebellion. We should recover its historical meaning. It is said that the founder, Je-U Choi, received a heavenly mandate in the event of interlocution with Sangjenim in the year 1860. The deep-rooted cosmology of Baedal people is evidenced by the historical fact that as many as six million believers of Donghak and its family religions chanted the following missionary message of Dongkyung Daejeon (동경대전) during the Japanese Occupation:

Si-cheon-ju Jo-wha-jeong Yeong-se-bul-mang-man-sa-ji

The idea of Donghak is well expressed by such statements in *Yongdam Yusa: The Hymns of Dragon Pool* as "The fate of the mysterious disease spreading across the entire world—is this not the return of gaebyeok?" and "Cultivate the Supreme Dao of Mugeuk, and the destiny of the next fifty thousand years shall be yours." These clearly reveal the aspiration of people at that time to cultivate a new world.

Je U Choi

Great universe unmasked as late as 100 years ago

We should be careful when applying our everyday perception of time to vast stretches of time across the universe. For example, in many films, primitive humans fight against dinosaurs. This is simply wrong. In fact, dinosaurs flourished during the Mesozoic era, while human beings first appeared in the Cenozoic era. Accordingly, human beings had no chance to meet dinosaurs. However, our vague and hazy perception of the past makes us imagine fights between humans and dinosaurs.

Our everyday perception of space also causes erroneous scenes when applied to the expanse of outer space. TV programs often show stars passing by like trees lining the street. This is another nonsensical depiction. Stars are at least a few light years apart from each other, meaning it takes a few years for a rocket moving at the speed of light to travel from one star to another. Therefore, spaceships would pass through black, empty space for thousands of years between stars.

"This is Earth HQ. Mars probe, please respond."
"This is the probe on Mars."

The Solar system is so inconceivably vast that the seemingly real-time communication described above is also impossible. Communication is performed by radio wave at the speed of light. The speed of light is 300,000 km per second, or 7.5 revolutions around the earth. It takes about one second for light to travel from Earth to the moon. However, it takes about four minutes to Mars when the planet is closest to Earth. Therefore, we should pay extra attention to the concepts of time and space when describing the universe.

Our galaxy travels across the night sky in summer and winter. In the East, the galaxy is called Eunhasoo (은하수), meaning a stream of glittering silver. Its nickname in the Western tradition is the Milky Way, indicating the milk of the Goddess Hera. The term "galaxy" also means milk. As I mentioned above, Herschel revealed that the Milky Way is a collection of countless stars. He expanded the imagination of humankind beyond the solar system to the galaxy. Unfortunately, Herschel believed that the solar system was at the center of the galaxy. This is another version of geocentricism.

Our galaxy has the shape of a disk that is thicker at the center than at the edges, and has a diameter of over 100,000 light years. The Milky Way visible at night is a cross section of this shape. Why does the disk-shaped galaxy look like a long stripe? In Figure 7-7, an ant at Point A, which is at the center of a hole in the disk, cannot see the disk's entire shape. It can only see the slashed area (that is, a cross section of the disk) as a revolving stripe. This is the reason why our galaxy looks like the Milky Way.

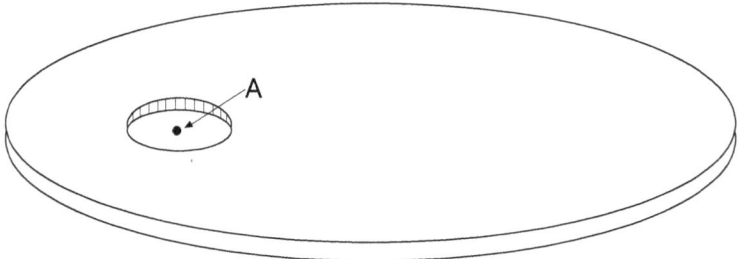

Figure 7-7 The reason why our galaxy looks like a stripe to us

Figure 7-8 is an image of our galaxy. It takes about 30,000 light years to reach the center of the Milky Way from the solar system in the direction of Sagittarius. Traveling in the opposite direction, that is, toward Taurus, it would take 20,000 light years to exit the galaxy. Earth and other planets of the solar system are revolving around the sun, while tilting with respect to the galactic plane.

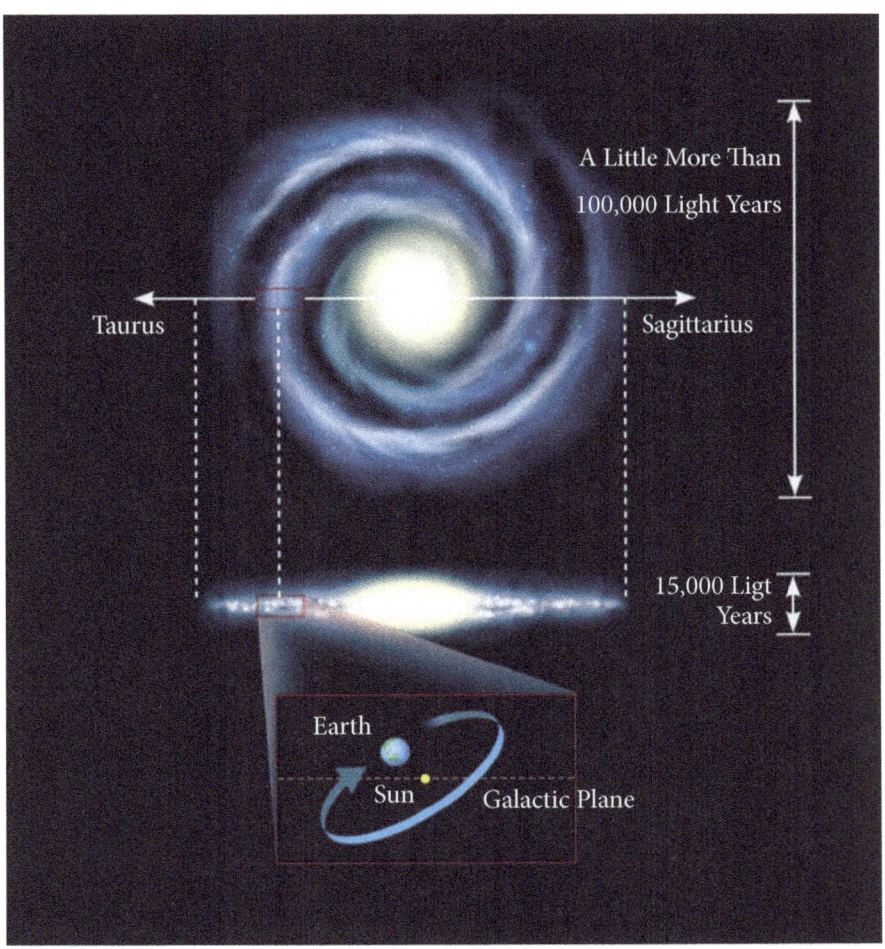

Figure 7-8 Structure of our galaxy

As you see in Figure 7-8, the galaxies are clearly thickest at their centers. Accordingly, the Milky Way looks thickest near Sagittarius, giving it the appearance of a snake that has just swallowed its prey. In Korea, Sagittarius appears in the summer. For this reason, the Milky Way looks most radiant in the summer in the northern hemisphere.

Figure 7-9 is an image of the Milky Way facing Sagittarius. The center of our galaxy is about 30,000 light years behind this strip of the Milky Way. In Figure 7-9, the red pimple-like spots are bright nebulae, and the black parts at the center of the Milky Way are dark nebulae. The yellow line indicates the South Dipper, which can be observed even by amateur astronomers.

Figure 7-9 Sagittarius Milky Way and South Dipper
(provided by Yeong-Bum Jeon)

Shapley was the first to discover that our solar system is at the edge of the Milky Way. He analyzed the observation that over a third of all star clusters are crowded near the Sagittarius Milky Way, and finally concluded that this area is the center of our galaxy.

In Figure 7-10, our galaxy looks like a whirlpool or a typhoon. How did astronomers figure this out? The shape of our galaxy is evident by looking at other galaxies. Imagine that we are trapped in an apartment. If we want to know the shape of the apartment building, we should look at a neighboring apartment building through the window. Figure 7-10 is an image of another galaxy, that is, a neighboring galaxy apartment seen from our galaxy apartment. After analyzing numerous observations, we have inferred that our galaxy has a similar appearance.

Figure 7-10 Face of galaxy (provided by Yeong-Bum Jeon)

Like our galaxy, other galaxies are enormous collections of stars with a diameter of tens of thousands of light years. Galaxies are millions of light years away from each other. For example, the Andromeda Galaxy, which may be our closest neighbor, contains about twice as many stars as the Milky Way and is about two million light years away.

Figure 7-11 Cluster of galaxies

Today, we know that the universe has a sponge-like structure with enormous holes and contains more than 100 billion galaxies. The universe has dark spaces where galaxies are sparse, as well as large bands of galaxies. A cluster of galaxies is a collection of such bands.

In 1929 Hubble discovered that the universe is expanding. He concluded that galaxies are withdrawing from us in every direction, and that the galaxy's speed of withdrawal is proportional to its distance from us.

It is noteworthy that our galaxy is not at the center of Hubble's expanding universe. The universe can be compared to an inflating balloon on which points move farther away from each other. From any point on the surface of the balloon, the other points move away faster as they become more distant, regardless of direction. This is the basic theory of the expanding universe.

Hubble

Cosmic God and His Disciples: Making the Expanding Universe

Earth god: "Master, Hubble has started observing outer space."
Cosmic god: "He has? Hurry and inflate it!"
Galaxy god: 'My black stones are all dead.'

Earth god: 'Do I have to do this every time Hubble gets near a telescope?'
Hubble: "The universe is expanding!"

Chapter 8

Theory of Relativity Changing Our Concepts of Time and Space

Einstein entered the academic world of physics when he published the special theory of relativity in 1905. It was just the prologue. Einstein's greatest contribution to physics came a decade later, when he published the general theory of relativity in 1915. While special relativity concerns time and space, general relativity is a theory of time, space and gravity. This theory can be expressed by a mathematically complex equation.

Space contraction and time dilation

Einstein published two theories of relativity: the special theory of relativity and the general theory of relativity. You are not far off the mark if you understand that the former is an easier theory applied to "special" cases while the latter is more difficult because of its "general" application. In fact, the special theory of relativity was published in 1905. Ten years later, in 1915, Einstein presented the general theory of relativity.

Einstein

For example, if we travel close to the speed of light in a horizontal direction, objects should contract and vertically lengthen. As if by magic, we move by contracting space; the space moving relative to us contracts. The point here is relative motion, which occurs regardless of whether we move or the space does. If a bus passes by me close to the speed of light, it will contract and reduce in length. From the viewpoint of a bus passenger, though, I am the thing in motion and become thinner.

When space contracts, time naturally changes. If we travel across space by contracting distance, our time cannot be the same as others' time. To sum up, in such a case, a person in relative motion experiences dilated time. The conclusions of special relativity can be summarized as follows:

① The space of an observer in relative motion is contracted in the direction of that motion.
② The time of an observer in relative motion is dilated.

The average lifetime of a particle called a muon is only about 0.000002 seconds. Produced by cosmic rays in the upper atmosphere, muons fall towards the surface of Earth at nearly the speed of light. Even if it falls at the speed of light, a muon's short lifetime is expected to limit its travel distance to 0.6 km (= 300000km/sec × 0.000002 sec). In reality, the particles are observed on the surface of Earth. Special relativity can explain this as follows.

As the muon falls close to the speed of light, its space is contracted according to the conclusion ① of special relativity, and thus it can reach the surface of Earth despite its short lifetime. From the viewpoint of an observer standing on the ground, it is reasonable to analyze, based on the conclusion ② above, that the particle reached the ground because its time was dilated and its life was extended.

For a more complete understanding, let's consider the famous pole-barn paradox. As shown in Figure 8-1, let's assume that there is a barn 10 meters long with front and back doors. Person "A" runs through the barn at close to the speed of light while carrying a pole 10 meters long. In the meantime, person "B" observes A. When A comes in the front door and goes out the back door carrying the pole, he is in relative motion from B's viewpoint. For this reason, the length of the pole should become shorter than 10 meters. In this case, from B's viewpoint, the pole can fit momentarily within the barn because it is shorter than 10 meters. For instance, if the pole is contracted to eight meters, it can fit in the barn momentarily by closing both doors simultaneously.

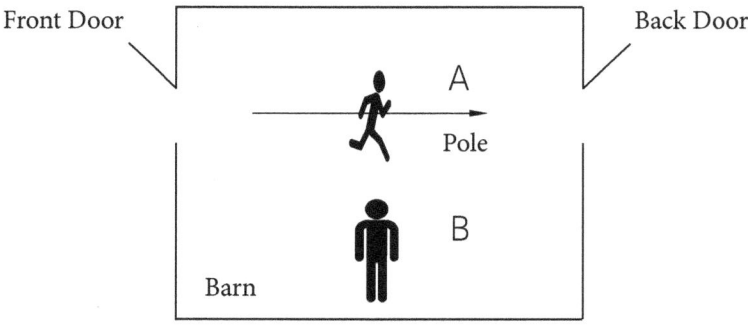

Figure 8-1 Lengths of pole and barn from B's viewpoint

From A's viewpoint, on the other hand, the length of the barn should be reduced, as illustrated in Figure 8-2. From A's viewpoint, if the pole is 10 meters long and the barn is eight meters long, then the barn can never hold the entire pole. However, according to special relativity, both Figure 8-1 and Figure 8-2 should be right. What is the solution?

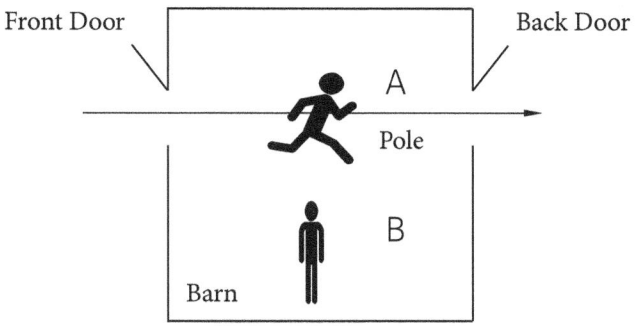

Figure 8-2 Lengths of pole and barn from A's viewpoint

The trick lies in the word "simultaneously." We have to divide Figure 8-2 into two cases, as shown in Figure 8-3. From B's viewpoint, he would argue that the doors are closed simultaneously. However, from A's viewpoint, the back door is closed as the pole passes through the front door, and the front door is closed as the pole passes through the back door. In other words, to A, the two doors are never closed at the same time.

Hence, A and B have different times, and so A exits the barn carrying the pole with no contradictions. If a bus passes in front of me at close to the speed of light, my flow of time becomes different from that of a bus passenger. According to the theory of relativity, time and space are not independent of each other but always change together; thus, they are collectively called spacetime. They are also called four-dimensional spacetime, meaning the combination of three-dimensional space and one-dimensional time.

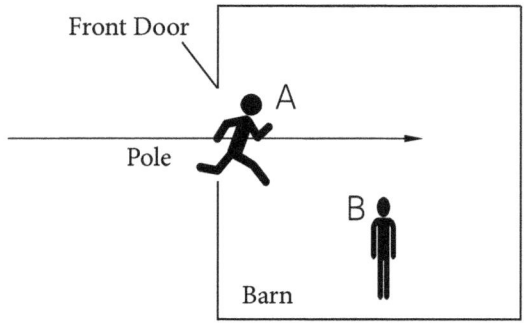

(A) Pole enters through front door

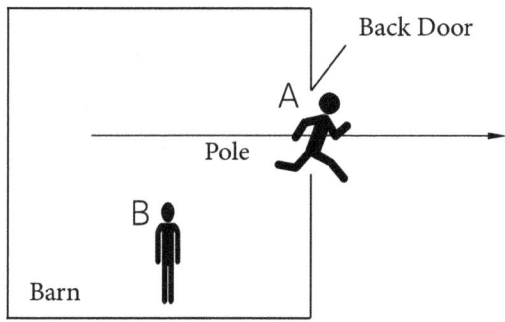

(B) Pole exits through back door

Figure 8-3 Pole and barn from A's viewpoint
when B closes the doors simultaneously

If special relativity is right, then the speed of light (300,000 km/sec) should be the fastest imaginable speed in the universe. For example, suppose we are in a rocket flying to the left at 90% the speed of light. If we see another rocket flying at 80% the speed of light in the opposite direction, that is, to the right, the latter rocket does not seem to fly at 170% the speed of light. In this case, the relative speed of the rocket is 99% the speed of light. If we ever discover a faster-than-light particle, the theory of relativity would be completely overhauled.

The famous equation $E=mc^2$ is derived from the above-mentioned conclusions ① and ②. Here, E, m and c represent energy, mass and the speed of light, respectively. The equation tells that energy can be converted to mass and mass to energy. Moreover, even a little mass can have a massive amount of energy since it is multiplied by the square of the speed of light. This is the very basic principle according to which humans are making nuclear bombs and hydrogen bombs.

Physics of curved time and space

While the special theory of relativity deals with time and space, the general theory of relativity, which was published by Einstein in 1915, addresses the triad of time, space and gravity. Many scholars contributed to the discovery of the former theory. However, Einstein completed the latter theory almost on his own, which still makes people marvel at his genius.

Let me introduce you to the famous Einstein's elevator. As illustrated in Figure 8-4, there are two elevators P and Q. P is fixed in a planet's field of gravity (or hangs from a rope suspended in midair), and Q is in outer space with no field of gravity and moves upwards with an acceleration equal to the acceleration of gravity acting on P. Strictly speaking, upward movement sounds nonsensical, but for convenience's sake, let's assume that Q is accelerating upwards.

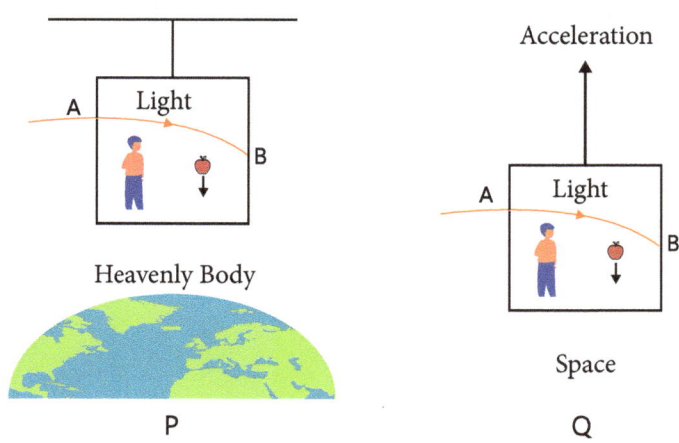

Figure 8-4 Einstein's elevators

In this case, the observers inside P and Q should describe exactly the same physics. This is Einstein's equivalence principle. For example, if the two observers drop apples, the two apples should both free-fall to the ground. As Q has accelerated motion, a beam of light entering laterally through hole A should bend and hit point B on the opposite wall, as shown in Figure 8-4.

Of course, the curvature of light is extremely small, but the light should bend anyway as long as its speed is not infinite. According to the equivalence principle, a beam of light which enters laterally through hole A in P should bend and hit point B on the opposite wall. If we consider only the case of P, the planet's field of gravity is the only factor that makes the beam of light bend. In other words, as illustrated in Figure 8-4, we have no choice but to conclude that light is bent because of the field of gravity.

Here, let's imagine that elevator Q stops, as illustrated in Figure 8-5. In this case, the observer inside the elevator feels weightless. If he puts an apple in the air, it does not fall. The observer also floats in midair, and there is no distinction between up and down. Moreover, the beam of

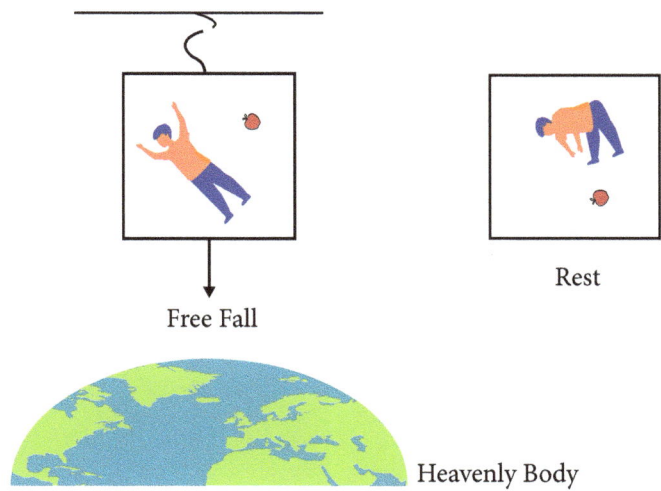

Figure 8-5 Einstein's elevator cut from the rope

light does not have a curved path in this case. Then, how can we make the observer inside elevator P experience the same conditions? The answer is simple: we just need to cut the rope and let the elevator free-fall down to the planet!

Compared with Newton's theory of gravity, the most remarkable idea of general relativity is that mass bends spacetime, generating a field of gravity. When traveling in bent spacetime, light naturally has a curved path. While light travels the shortest path between two points, its path is not a straight line in curved spacetime. Accordingly, mathematically speaking, light is curved in non-Euclidean space.

According to Newton's theory of gravity, an object falls to a heavenly body because it is caught in the heavenly body's gravitational pull. However, the theory of relativity interprets the fall as the result of the object's motion in spacetime curved by the gravity of the heavenly body. For example, if we put a heavy ball (heavenly body) on a thin rubber sheet, the rubber sheet will curve as illustrated in Figure 8-6. Moreover, if we put a small, light ball on the rubber sheet already depressed by the heavy ball, the small ball will roll around and fall inward towards the heavy ball, as shown in Figure 8-6.

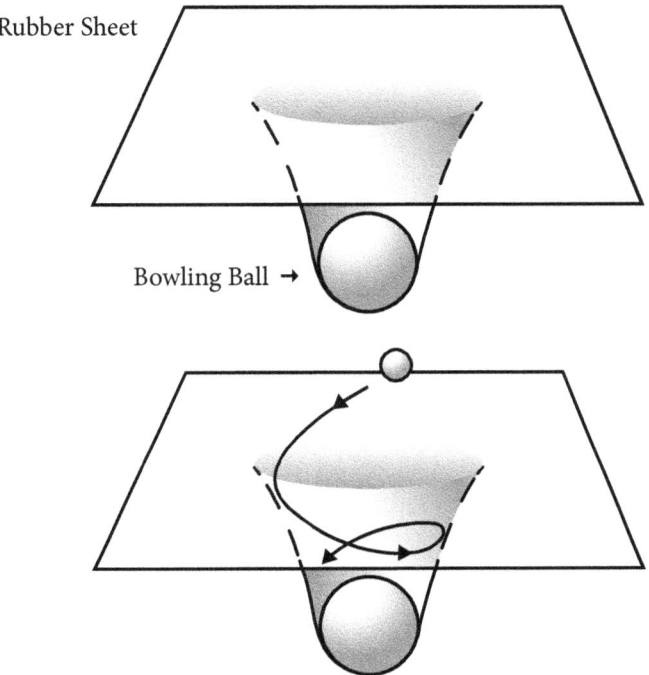

Figure 8-6 Gravity as curved spacetime

Black hole, the ugly duckling

The curved spacetime of general relativity can be described by Einstein's gravitational field equations, which was published in 1915. The following year, Einstein received a letter from Schwarzschild in Germany saying that he had solved the equation. According to Schwarzschild's solution, a beam of light should be curved at an angle as small as approximately 2 seconds near the sun, as shown in Figure 8-7. Recall that 1 degree is equivalent to 60 minutes or 3,600 seconds, indicating how small this curvature is. During a solar eclipse in 1919 in Africa, a team led by Eddington confirmed Einstein's equation. The world was shocked by the news.

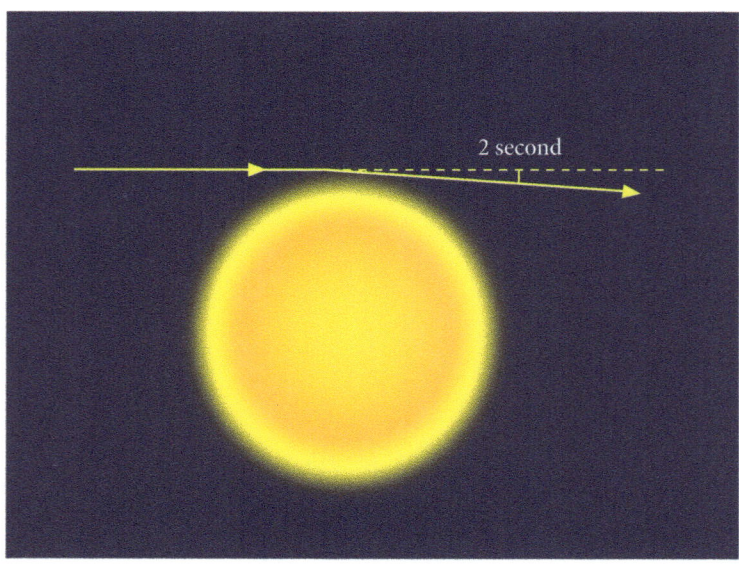

Figure 8-7 Path of light curved near the sun during eclipse

Thus, general relativity has been accepted as a more rigorous theory that can correct the Newtonian theory of gravity. I say "correct" because although Newton's theory is as reasonable as general relativity when discussing weak gravity, it becomes a mere approximation of general relativity under conditions of strong gravity. Moreover, there are general relativity effects of strong gravity that Newton's theory cannot explain.

In Figure 8-7, if the sun shrinks in size without losing mass, the surface gravity becomes stronger, which increases the curvature of light. Schwarzschild argued that if the radius of the sun became as small as 3 km, light would be absorbed. This is because the sun's gravity would become stronger if the sun shrank in size with no change in mass. If the gravity became so strong that it absorbed not only light around the sun but also light emitted from it, a black hole would form. If the ball in Figure 8-6 is so heavy that it makes a deep pit in spacetime, as illustrated in Figure 8-8, this would be a black hole. Nothing, not even light, can escape from a black hole's field of gravity.

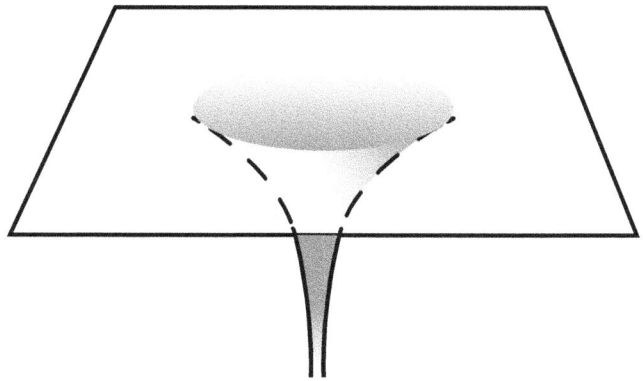

Figure 8-8 Black hole spacetime

Since not even light can escape, it looks black to us, hence the descriptor "black." Also, as everything is sucked into it, it is called a "hole." In this regard, a black hole's surface is called the event horizon. Strictly speaking, "horizontal plane of event" is a more appropriate term, but "event horizon" has already been established as a technical term.

On an event horizon, spacetime is sucked into the center of the black hole at the speed of light. As the spacetime "escalator" moves inward at the speed of light, light emits from the black hole in vain. Thus, nothing can escape from it.

The spacetime escalator becomes slower than the speed of light at a more peripheral point on an event horizon. When the speed of the escalator is zero, the effects of general relativity disappear. For instance, if a person riding an escalator moving inward on an event horizon throws an apple outside every second, another person standing outside the escalator may catch the apple every five seconds. That is, one second around the black hole may become five seconds outside it. This is the time-dilation effect caused by the strong gravity of the black hole.

Let's imagine one of a pair of twin brothers free-falls toward a black hole while the other is looking on. If he measures time by feeling his pulse, he would not see any change in the time interval. After a certain duration, he will reach the event horizon. From the other brother's viewpoint, however, as his brother moves closer to the black hole, his falling speed seems to decrease, as does his aging. Finally, when the brother reaches the event horizon, time seems to stand still. That is, during his lifetime, one of them can never see the other disappear into the surface of the black hole.

The sun's shrinkage into a star with a 3km radius can be compared to Earth's shrinkage into a planet with a 9mm radius. In other words, if the radius of the sun became 3km, Earth's radius would be 9 mm. Today, we know that such shrinkage actually occurs in the universe. At that time, however, most scientists did not accept it. They concluded that there were no black holes in space even if general relativity was right. The black hole became an ugly duckling, and scientists no longer had any interest in it.

Universe born with the Big Bang

In 1917, Einstein published a paper on cosmology based on general relativity. The paper applied the bending of spacetime by gravity to the entire universe. Einstein presented a cosmic model in which the entire spacetime of the universe is warped by the gravity of all the galaxies.

Einstein assumed a static universe that never evolves. When he proposed his theories, a dynamic expanding universe was unimaginable. The observations that had been accumulated so far showed such a static universe, which Einstein tried to describe theoretically.

Einstein also assumed that the galaxies were evenly distributed across the entire universe. Of course, he was aware of some irregularities like clusters of galaxies. Nevertheless, he believed the galaxies were distributed uniformly in every corner of the universe. As this assumption is not significantly mistaken in view of observations and is rather mathematically useful, it is still frequently used.

Thus, Einstein actually suggested that the universe was the surface of a huge four-dimensional sphere or, to be precise, a three-dimensional surface space. He thought the gravity of all the galaxies could make such a uniformly closed space, in which light leaving us travels across the warped space and comes back to us. This space is a finite universe with a constant volume.

Accordingly, Einstein inherited Newton's problems because the universe could not be static. Like the stars of the Newtonian universe, the galaxies in Einstein's universe attract but do not repel each other. Any static universe with a finite number of galaxies shrinks into a point due to gravity and then collapses. Newton solved this problem by assuming an infinite universe, which Einstein could not opt for.

Einstein's solution was a little awkward. He argued that there should be not only attraction or gravity but also repulsion between galaxies. In order to prevent the collapse of the universe, he introduced the idea of repulsion between galaxies that were attracting each other, as support.

Fortunately, Hubble discovered the expansion of the universe, which is not static but dynamic. In other words, if Einstein's static universe expanded, we would have the dynamic universe observed by Hubble. The dynamic cosmic model thus replaced the static one, and Einstein was relieved from this problem. After Hubble's discovery, Einstein reportedly regretted his awkward assumption of attraction as the biggest mistake of his life.

However, Einstein's gravitational field equations caused no problems in theorizing the expanding universe observed by Hubble, since his work intrinsically encompassed a solution to describing dynamic universes. Contemporary cosmologists including Friedmann, Lemaitre, Robertson and Walker conducted an in-depth study on the equation. Finally, three candidate solutions were proposed.

Figure 8-9 shows two of the three solutions. (A) is a closed four-dimensional spherical universe, and (B) is an open saddle-shaped universe. In both cases, spacetime is warped, and the sum of all the internal angles of a triangle is not 180 degrees. The sum is greater than 180 degrees in the spherical universe and less than 180 degrees in the saddle-shaped one. In addition, the spherical universe is closed and has finite volume, whereas the saddle-shaped universe is open and has infinite volume. The third candidate is a universe with a four-dimensional flat plane applying Euclidean geometry. Of course, in this universe, the sum of all the internal angles of a triangle is 180 degrees.

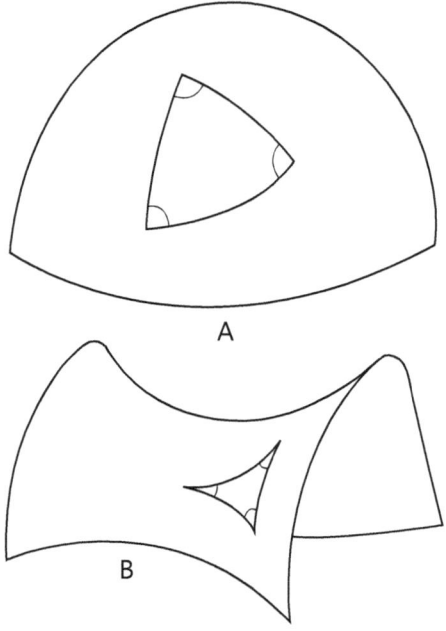

Figure 8-9 Four-dimensional universe

The idea of the expanding universe implies that if we wound back time, all the galaxies would gather at one ultimate point. If so, the universe in the beginning might have been unimaginably dense and hot. It is therefore quite natural to infer that the expanding universe was born in a massive explosion, i.e., the Big Bang.

There is another cosmology that does not accept the hot beginning of the universe, unlike the Big Bang (BB) cosmology. The basic idea is that the "horrible" Big Bang can be avoided if the galaxies are removed one by one as we wind back time. According to this cosmology, the galaxies are born one by one over the passage of time. This is the continuous creation (CC) cosmology. The CC universe has no change in shape over time. Having an unvarying shape with no beginning nor end, the CC cosmology is also called steady-state model.

The conflict between BB and CC cosmologies was one of the biggest issues in the history of science during the 1950s and 1960s. BB was mainly advocated by American cosmologists including Gamow, while CC was argued by British cosmologists, one of whom was Hoyle. BB seems to have won by decision. Cosmic microwave background radiation (CMBR, 우주배경복사) is the conclusive evidence. CMBR is comparable to steam lingering in a bathroom after a hot shower. It is the evidence that the universe was hot in the beginning.

Cosmic God and His Disciples: Making the Expanding Universe

Cosmic god: "Why is spacetime warped?"
Einstein: "Because of gravity!"
Earth god: 'He knows too much…'

Cosmic god: "You call yourself a god? He knows more than you!"
Earth god: "Ouch! He's Einstein!"
Einstein: 'Maybe I can be a god.'

Chapter 9

Quantum Physics: Bringing Light to Atomic Energy

Stars shine because light atoms combine to produce heavier atoms and release energy. With the development of theories on the structure and evolution of stars, the black hole ceased to be treated as the ugly duckling and became a beautiful swan. In quantum physics, particles are waves, and a vacuum is not empty. This revolutionary idea has not only resulted in the inflation cosmology but also given clues to the beginning and end of time.

Nuclear fusion, the powerhouse of stars

Figure 9-1 is a typical representation of hydrogen, the simplest atom. In this model, a negatively charged electron with light mass orbits around a proton, which is a positively charged particle. But why is the electron captured by the proton? Because there is electromagnetic attraction between positive and negative charges. This is to say that the electromagnetic force is the very thing causing the proton and the electron to constitute the hydrogen atom. The electromagnetic force becomes an attractive force between charges with different signs, and a repulsive force between charges with the same sign.

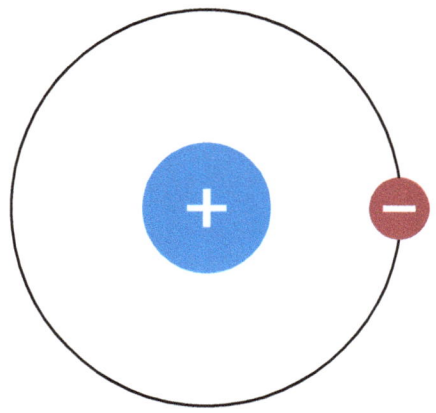

Figure 9-1 Structure of hydrogen atom

If the proton has a charge of +1, the electron has a charge of exactly -1. Thus, the hydrogen atom has a charge of 0 (= +1-1). In other words, the hydrogen atom is electrically neutral. Then why, as seen in Figure 9-1, does the electron orbit around the proton, and not the other way round? This is because the proton is about 1,840 times heavier than the electron in terms of mass. If the electron were much heavier than the proton, the latter would orbit around the former.

A helium atom has two protons and two neutrons in its nucleus. Accordingly, a helium nucleus has a charge of +2 and is about four times heavier than a hydrogen nucleus. For this reason, a helium atom can have two electrons, and its total charge is +2 -2 = 0. If one electron leaves a helium atom, the helium atom will have a charge of +1.

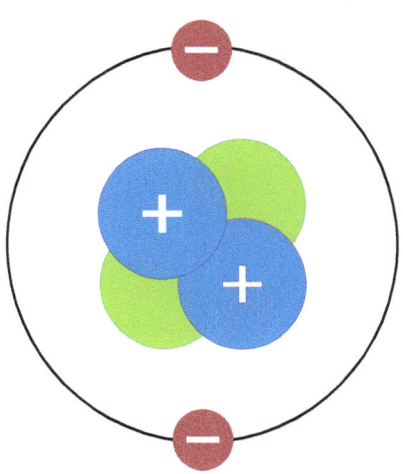

Figure 9-2 Structure of helium atom

Approximately 75% of the mass of the visible universe is in the form of hydrogen. Helium makes up about 25% of the mass. This is good news to astronomers because hydrogen and helium atoms have the simplest structures and we have a good understanding of them.

Note that the two positively charged protons in Figure 9-2 do not push each other away but stay very close. Why do they not move apart? The protons are actually pushing each other with electromagnetic force. Nevertheless, a far stronger force keeps them close to each other. This is the nuclear force, which is 137 times stronger than electromagnetism.

Figure 9-2 helps us understand one important feature of nuclear force. A nuclear force acts over very small distances, such as within the diameter of a nucleus. When a helium nucleus consisting of two protons and two neutrons is made using four hydrogen nuclei or four protons, as shown in Figure 9-3, energy is released. This is because the difference between the mass of four protons and that of one helium nucleus is converted to energy according to the formula $E=mc^2$. This is the principle of the hydrogen bomb.

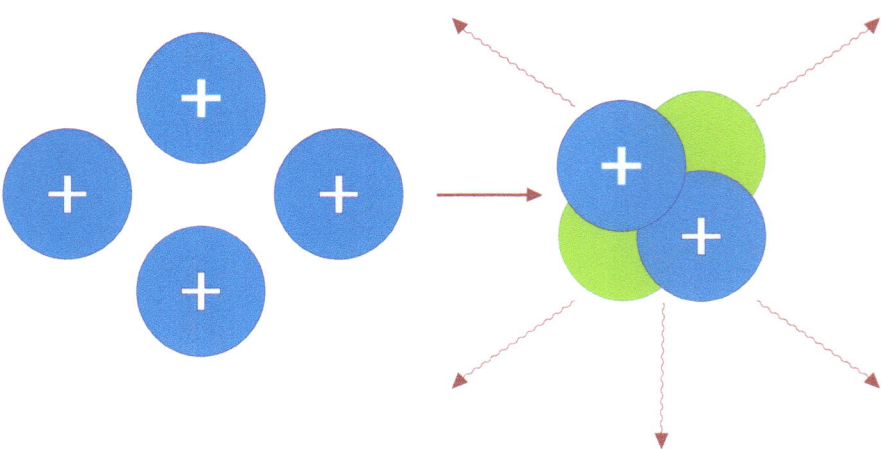

Figure 9-3 Hydrogen nuclear fusion

When the nuclei of low-mass elements like hydrogen and helium are combined into a larger nucleus, energy is released. This phenomenon is called nuclear fusion. Under the right conditions, nuclear fusion can proceed to produce an iron nucleus, which is the most stable in the universe. On the other hand, when the nucleus of a high-mass element like uranium is split, energy is also released. This is nuclear fission. Nuclear fusion and nuclear fission are both nuclear reactions. Nuclear plants produce electricity using nuclear fission, while nuclear fusion powers stars in their birth and evolution.

Let's imagine a box full of hydrogen atoms in outer space at an extremely low temperature, as illustrated in Figure 9-4. No matter how long we wait, those randomly moving hydrogen atoms will not collide with each other and become helium atoms.

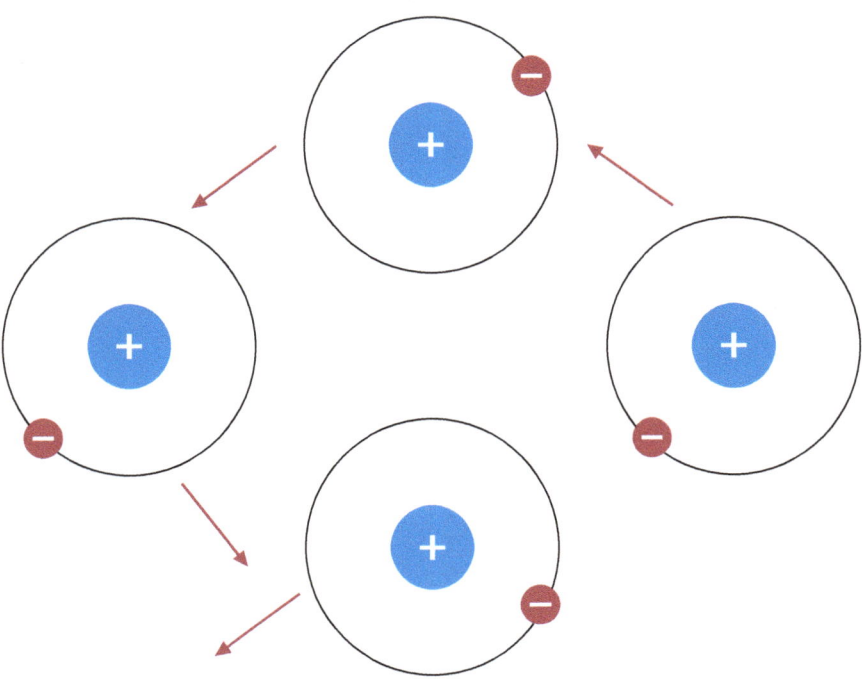

Figure 9-4 Hydrogen atoms in free motion

However, if the temperature rises, the hydrogen atoms move more quickly. In outer space, the hydrogen atoms will move at a sufficient speed when the temperature has increased to about 10,000°C. They will then collide with each other, and the protons will release electrons. As a result, the box will become full of freely moving protons and electrons. This is illustrated in Figure 9-5. This state in which each atom is ionized is called plasma.

As the visible universe is mostly composed of hydrogen, the temperature of 10,000°C is very significant in astronomy. Nevertheless, the state depicted in Figure 9-5 is not sufficient to make a helium atom.

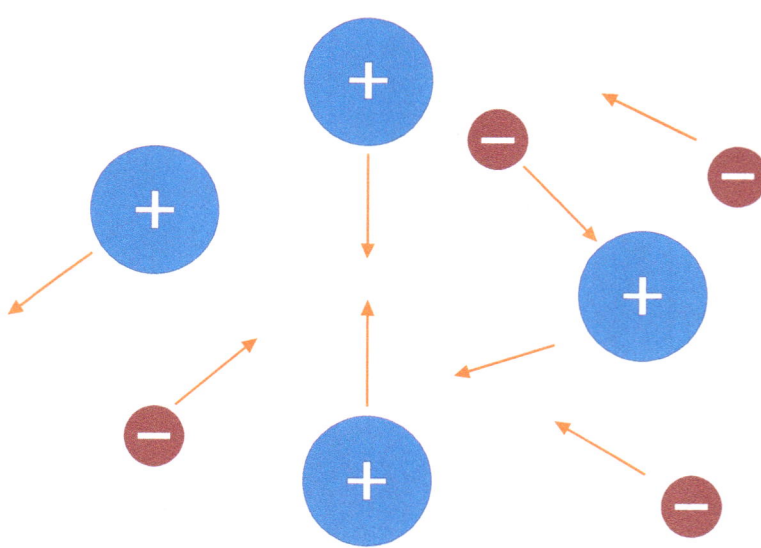

Figure 9-5 Ionized hydrogen atoms

This is because, as shown in Figure 9-6, two protons experience larger repulsion as they approach each other. When they get too close, a strong repulsive force keeps them away from each other. In that case, is it impossible to make helium atoms no matter how much the temperature is raised? Without nuclear force, no helium atoms could be made.

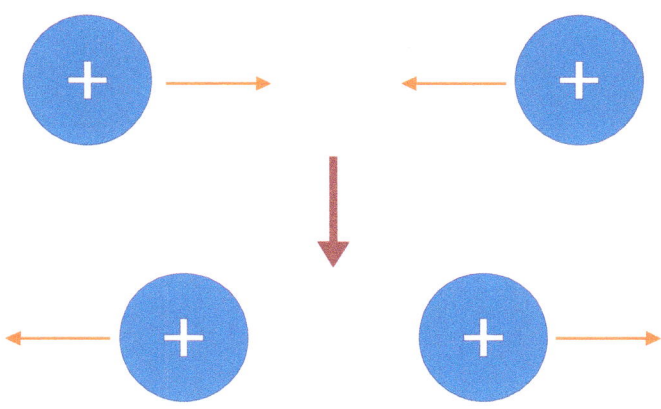

Figure 9-6 Two protons approaching each other

As mentioned above, nuclear force begins to act when protons get as close together as they do in the nucleus of an atom. Two protons approaching each other stick together when they are within the distance in which nuclear force is dominant.

That is, if the temperature is sufficiently high, two protons can move fast enough and get close enough to overcome electromagnetic repulsion and become fused together by nuclear force. That temperature is approximately 10 million degrees Celsius. At this temperature, hydrogen atoms in a plasma state begin to form helium nuclei. This process of nuclear fusion is accompanied by the release of energy.

Black hole becomes swan

Stars are born through the gravitational contraction of interstellar matter that is mostly composed of hydrogen atoms. As interstellar matter moves in faster spirals and undergoes gravitational contraction into the center, the internal temperature gradually increases. As mentioned above, when the temperature reaches 10,000°C, most hydrogen atoms are ionized. When the temperature at the center reaches 10 million Celsius, energy begins to be released due to nuclear fusion where hydrogen atoms form helium. A star begins to shine.

Helium nuclei have a relatively large mass, which causes them to sink to the center of the star. Thus, the stars have a structure as described in Figure 9-7. In this case, helium is nothing but ash produced by nuclear fusion. Please note that Figure 9-7 exaggerates the accumulation of helium.

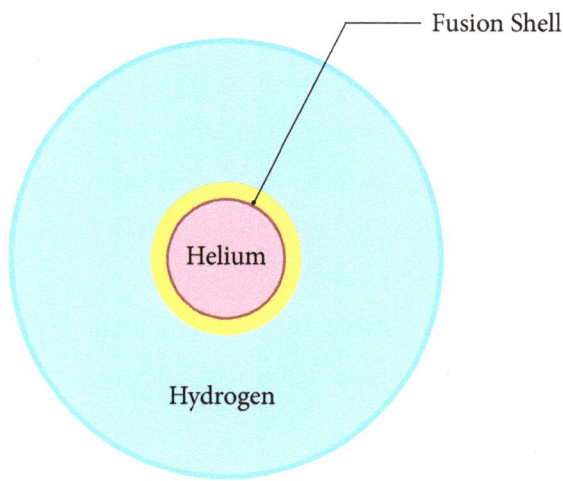

Figure 9-7 Structure of star

When the temperature at the star's center where helium is accumulated exceeds far beyond 100 million Celsius, another process of nuclear fusion making carbon from helium is triggered. This reaction, using helium as fuel, is the principle behind the "helium bomb." Now, helium is no longer ash but becomes fuel in the nuclear reactor at the center of the star.

This process produces high-mass elements like carbon, nitrogen and oxygen. Of course, it occurs only in massive stars. The largest stars additionally produce elements such as neon, magnesium and iron, giving them an onion-like structure shown in Figure 9-8.

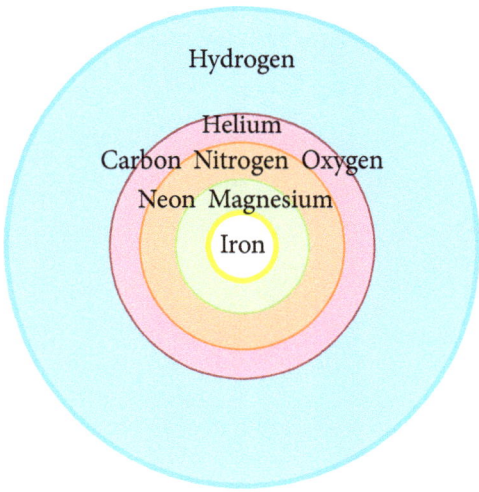

Figure 9-8 Structure of heavy star

Stars in their old age shrink due to a decrease in nuclear fusion energy. Stars 10 to 30 times more massive than the sun become neutron stars at their center. A neutron star with similar mass to the sun has a radius as short as 10 km. If it shrinks a little further, it becomes nothing but a black hole. In fact, it was discovered that stars 30 times more massive than the sun leave black holes behind.

With this discovery in the 1960s, the black hole transformed from an ugly duckling to a beautiful swan. The name "black hole" was coined by Wheeler in 1969. Before that, black holes were called "frozen stars" or "collapsed stars." Wheeler's renaming symbolizes the dramatic change of luck for those stars. The mass of a black hole has no limit. Theoretically, there may be many types of black holes, with mass ranging from 1/100,000 g to infinity.

In 1963, nearly 50 years after Schwarzschild solved Einstein's gravitational field equations, Kerr proposed a solution for rotating black holes. Schwarzschild's solution explains only black holes that do not spin but stand still. Accordingly, in astronomy, the terms "Schwarzschild black hole" and "Kerr black hole" refer to a static black hole and a rotating black hole, respectively.

Until 1960s, "the corpse of a star" was the safe answer to the question "What is a black hole?" However, this response has become inaccurate, as recent discoveries show black holes over one million times more massive than the sun at the center of many galaxies. Now, we have a common understanding that most galaxies have a huge black hole at their center.

In Figure 9-9, a wormhole connects two black holes as described by Figure 8-8. The wormhole is a passage created by gravity. This term originates from the idea that the shortest way for a worm to traverse two opposite points on the surface of an apple is through a hole. Since Newton, the apple has been frequently referenced to explain gravity. Now, astronomy features a worm in the apple.

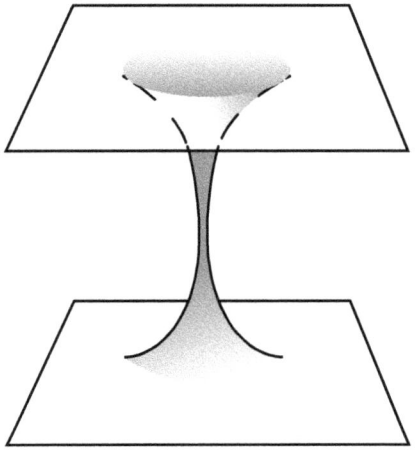

Figure 9-9 Wormhole

In the theory of relativity, a wormhole was a passage connecting two black holes. However, there was a problem. Even if one entered a black hole and survived the trip to the other one at the opposite side, he could not escape from it. Thus, a white hole seemed necessary to regurgitate everything ingested by the black hole. This was the invention of science fiction writers. Thus, black holes and wormholes are scientifically justified, while white holes are still an imaginary phenomenon.

Matter and vacuum

We know there are three fundamentals in the universe: gravity, electromagnetic force and nuclear force. The last one can be further divided into two types, but we do not need to explore that division here. Gravity acts between masses and is the weakest force in the universe. Accordingly, gravitational force is negligible when considering atoms. If, for instance, a hydrogen atom were made by a gravitational force between a proton and an electron, the atom would be several light years in radius, which is unimaginable.

Nevertheless, the main influential factors for the structure of the universe are gravity and electromagnetic force, rather than the strongest nuclear force. This is because both gravity and the electromagnetic force are influential over infinite distances, while the nuclear force can act only in a space as small as an atomic nucleus.

In fact, compared to gravity, electromagnetic force plays a minor role in determining the structure of the universe since celestial bodies themselves are not electrically charged. For example, we cannot say that a star is positively charged. Planets orbiting a star constitute a group like our solar system, stars make up a galaxy, and galaxies are members of a cluster. Thus, the entire cosmos is constructed under the absolute influence of gravity.

In the figures above, I described protons and electrons as if they had the shape of a billiard ball. Though certainly convenient and useful for explaining those particles, it is far from reality. According to modern physics, a particle is a quantum with wave properties. The concept of quantum was first introduced in 1901 by Max Planck.

Planck

Whether matter is a particle or a wave in essence was discussed as fiercely as the controversy in astronomy between BB and CC. Modern physics accepts particle-wave duality. As electrons actually have wave properties, the illustration of a hydrogen atom in Figure 9-1 is inaccurate, since the electron is described to have an orbital motion like Earth revolving around the sun. In the hydrogen atom, the electron is somewhere near the radius as shown in Figure 9-10. We can only say with the highest probability that the electron is located on the radius.

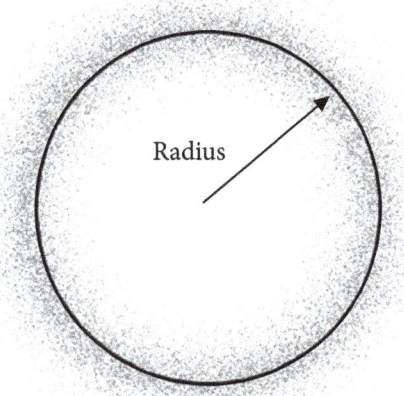

Figure 9-10 Location of the electron in a hydrogen atom

Light is both a particle, called a photon, and a wave, or an electromagnetic wave. Einstein's famous equation $E=mc^2$, which was derived from his special theory of relativity, clearly reveals the relationship between light and matter. According to the equation, energy can be converted to mass, and mass to energy. For example, a photon can be changed to another particle like an electron. However, in this case, as shown in Figure 9-11, not only an electron but also a positron, which is the antiparticle of the electron, is generated, which is called "pair production."

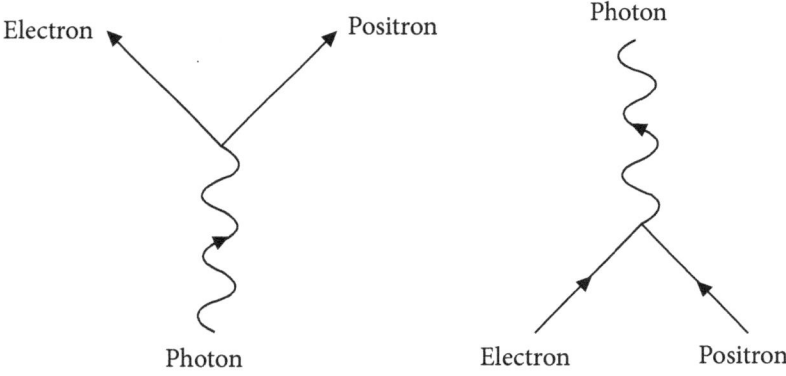

Figure 9-11 Pair production and pair annihilation

An antiparticle is a twin of a particle, with the same mass but different charge. For example, the antiparticle of an electron with a negative charge is a positron with a positive charge. The pair annihilation between a particle and its antiparticle produces a photon. According to modern physics, a vacuum is a space in which both pair production and pair annihilation of particles and antiparticles constantly occur. In other words, a vacuum is not a space with zero energy.

Based on pair production and pair annihilation, Hawking announced in 1974 that black holes can emit light like other celestial bodies. This dramatically changed scientists' attitude toward black holes. Accordingly, it is widely agreed that black holes are not monstrous bodies as assumed before. Furthermore, Hawking also discovered that black holes are not always so greedy as to absorb everything but are sometimes kind enough to share it back.

Hawking expressed his discovery with the statement "Black holes ain't so black". This was a turning point in the study of black holes. Currently, they are recognized as energy tanks.

In Figure 9-12, both particles that are pair-produced near a black hole can be absorbed into the event horizon, or only one of them may be absorbed. The latter is the more interesting case because from the viewpoint of an observer standing outside, the black hole seems to suddenly create a particle. We can conclude that such a phenomenon constantly occurring over the whole surface of the black hole is nothing but the emittance of light from the black hole. This emitted light is called Hawking radiation.

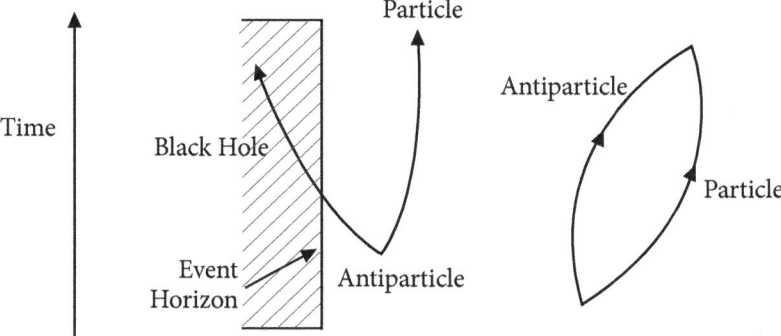

Figure 9-12 Pair production and pair annihilation near a black hole

Hawking radiation comes from a black hole. For this reason, the black hole itself may gradually lose mass and evaporate. The lower the mass, the more remarkable the evaporation of the black hole. According to Hawking, the temperature of a black hole is inversely proportional to its mass. Naturally, a very small black hole evaporates in a process similar to an explosion. Thus, such a small black hole is actually the same as the white hole mentioned above.

The lower its mass, the shorter time the black hole lasts. At last, like other celestial bodies, the surface temperatures of black holes can be defined, and electromagnetic fields can be formed on the surface. We now have new areas of physics like black hole thermodynamics and black hole electrodynamics. The black hole was born as an ugly duckling, grew into a swan, and now flies high like a phoenix.

The beginning and the end

Approximately 75% of the mass of the visible universe is in the form of hydrogen. Helium makes up about 25% of that. The initial temperature of the universe must be over 10 million degrees Celsius to create helium through nuclear fusion. Accordingly, the high proportion of helium is the evidence that the cosmos began at an enormously high temperature. Together with the CMBR, the nearly 25% proportion of helium among the constituent elements of the universe made a decisive contribution to the victory of BB over CC.

According to the BB model, about three minutes after the universe was born 15 billion years ago, the temperature dropped below 10 million Celsius and nuclear fusion stopped. The amount of helium produced during these three minutes accounts for about a quarter of all the space matter known to us. This theoretical estimation comports well with observable evidence. Along with CMBR, the existence of helium was the conclusive evidence that led BB to prevail over CC.

The BB model states that the temperature of the universe dropped to below 1,000°C about 300,000 years after the Big Bang. If so, all the electrons were captured by hydrogen or helium nuclei. Accordingly, photons, which had been constrained by electrons, came to move freely.

The light that began to spread from that moment is the CMBR we observe today. Strictly speaking, CMBR is not the light that began to spread at the moment of the Big Bang. In 1964, Penzias and Wilson accidently discovered the CMBR cooled down to -270°C. The discovery earned them a Nobel Prize.

The cosmic expansion that started with the Big Bang is slowing down due to the constant interruption of the gravity of galaxies. If the residual energy from the initial intensity of the Big Bang drops below a certain criterion, the universe will gradually stop expanding and ultimately stand still. In other words, the gravity of galaxies may stop the expansion of the universe. After that, the universe will start shrinking again, in the opposite phase of the Big Bang called the Big Crunch. If the initial intensity of the Big Bang was high enough, however, the expansion of the universe may decelerate but will never stop.

The destiny of the universe depends solely on the quantity of matter in space. If density exceeds a certain value, expansion is strongly interrupted by gravity and the universe shrinks again. Otherwise, that is, if density is below a certain value, expansion is not interrupted and the universe expands forever. The boundary between these two scenarios is the threshold density, which corresponds to about 0.00…0045 (30 zeros)!

However, we can observe only 5% of the value through stars and galaxies. Thus, we have no choice but to believe that a considerable portion of cosmic matter is invisible to us. We refer to this as "dark matter." Identifying dark matter is one of the most important issues in modern physics. It is estimated that dark matter accounts for about 25% of the boundary value.

However, even considering dark matter, we know only 30% of the boundary value. What is the remaining 70% of matter filling the universe? Believe it or not, modern cosmology tells us that vacuum fills the remaining 70% of space. Here, a vacuum is not an empty space but full of energy from the viewpoint of quantum physics.

CMBR is almost uniform in all directions. Accordingly, we obtain the same information from observations in any direction. This is marvelous; as CMBR is radio frequency radiation, it reaches us at the speed of light. Guth proposed cosmic inflation as a solution to this mystery.

This is not related to rising cosmic prices but states that the universe abnormally expanded within one minute immediately after the beginning. The expansion started slowly, became inflationary, and then slowed down.

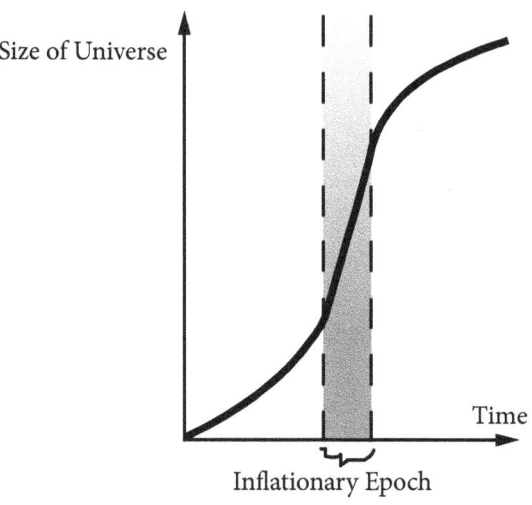

Figure 9-13 Inflation and the size of the universe

Before the inflation, immediately after the beginning, all matter was so small that it mixed together well. The inflation made the universe expand over 10^{30} times within one minute. As CMBR began to spread in this expanded universe, it became uniform throughout space.

The inflation ended in every part of the universe. However, what would happen if it continued in one corner of the universe? Astonishingly, such a case could bring forth a baby universe. It is similar to when we grasp an inflating balloon in our hands and a little sub-balloon forms between our fingers. In this regard, as shown in Figure 9-14, a baby universe should be connected to its mother universe through a wormhole, like an umbilical cord. Once the wormhole collapses later, the baby universe and the mother universe are separated from each other and become two parallel universes.

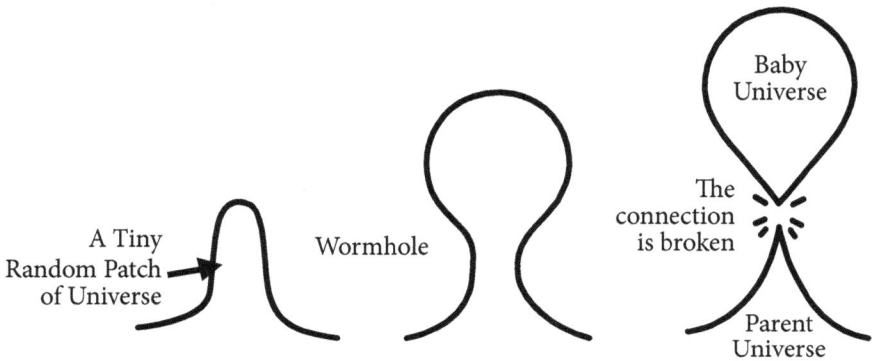

Figure 9-14 Birth of a baby universe

A baby universe can then bring forth another baby universe. If inflation proceeds, innumerable parallel universes will flourish like grandchildren and great-grandchildren within just one minute. This is the famous "bubble universe" of modern cosmology. Some universes with excessively high density were destroyed by the Big Crunch immediately after the Big Bang. We were lucky enough to be born in one of those countless parallel universes which did not immediately collapse.

According to quantum physics, when a universe changes phase from a high-energy to a low-energy vacuum, inflation occurs. The vacuum is not all the same. Phase transition is a physics term referring to a change of state in matter, such as water freezing into ice.

During the first billion years after the beginning of the universe, the first celestial bodies were born. Those celestial bodies include quasars.

The universe we live in is composed mostly of hydrogen and helium. However, the incessant nuclear fusions of stars will run out of hydrogen and helium in the future, after which no more stars will be born. Galaxies will not have any shining stars except corpses like white dwarves, neutron stars and black holes. This will happen after about one trillion years.

The universe will either keep expanding to infinity or stop expanding, shrink again and go back to its initial state. Both in an open and closed universe, if the actual density of matter is close to a threshold density, the life of the universe is significantly extended. If so, after 10^{27} years, each galaxy will have become a huge black hole. In other words, the huge black hole at the center of each galaxy will consume the stars that have already become corpses.

After about one billion years in this universe, the first stars were born. However, after about one trillion years, the birth of stars will cease, leaving only corpses of stars like white dwarves, neutron stars and black holes. After about 10^{27} years, every galaxy will have become a huge black hole. Likewise, after 10^{31} years, entire clusters of galaxies will become one enormous black hole.

This is the message conveyed by the last part of Cheonbugyeong (The Scripture of Heavenly Code).

One is the end; in Nothingness ends One (일종무종일).

Is it meaningful to consider something beyond the future? Modern cosmology seems to go beyond science with respect to ultimate time and space. Any cosmology, whether based on thoughts like in the East or mathematical equations like in the West, should depend fundamentally on spiritual judgements. We should not disregard Eastern cosmology as unscientific and blindly believe in Western cosmology.

That said, it is absurd to argue that the ancient East perfectly anticipated modern cosmology. For example, we need not say that the galaxy-like shape of Taegeuk demonstrates the knowledge of galaxies in the ancient East. As mentioned above, the structure of galaxies was discovered as recently as 100 years ago.

Nevertheless, there remains a possibility that the language of the East is used to illuminate Western cosmology. I said the emergence of Jeongyeok began the second season of Eastern cosmology. Jeongyeok, which is actively examined in many current studies, needs to be demonstrated from a modern perspective due to its numerous astronomical interpretations.

Any contents thus verified should be restated with modern scientific terms so that a new Eastern cosmology can be introduced. For instance, Jeongyeok describes the history of the universe in the order of Mugeuk, Taegeuk and Hwanggeuk. Using these terms, the modern cosmology may explain the beginning as Mugeuk, high-energy vacuum as Taegeuk, and low-energy vacuum as Hwanggeuk. We already know that the "nothingness (무)" of Mugeuk agrees with Cheonbugyeong.

The Beginning is described in the very first part of Cheonbugyeong:

One is the beginning; from Nothingness begins One (일시무시일).

Cosmic God and His Disciples: Big Bang Experiment

Cosmic god: "I have a bad feeling about this…"
Galaxy god: "Do you really want to do the Big Bang experiment?"
Earth god: 'I can't fully understand it without a test.'

Galaxy god: "I told you not to do it."

Epilogue

After years of study, you will naturally realize that Koreans, who believe in heavenly posterity, led the history of heaven. It is quite natural that Korea has the national flag of Taegeukgi, which is worthy of the cultural heritage of humanity.

The Emperor Gojong of the Chosun dynasty was so wise as to suggest that Taegeukgi, which incorporates Taeho Bokhui's eight trigrams in Figure E-1, be the national flag. As I mentioned above, Taegeukgi itself deserves to be celebrated as an excellent cultural heritage of humanity. It symbolizes the essence of Eastern cosmology. In 1882, the Emperor Gojong commanded Young-Hyo Park, the special envoy to Japan, to carry Taegeukgi, as seen in Figure E-1.

E-1 Taegeukgi with Bokhui's eight trigrams

On a ship to Japan, Young-Hyo Park devised another type of Taegeukgi including only four heavenly trigrams instead of all eight. As you see in Figure E-2, he removed the four earthly trigrams (Tae, Jin, Son and Gan) at the diagonal corners and left the four heavenly trigrams (Li, Gam, Geon and Gon) in the directions of East, West, South and North. This is the archetype of the current Taegeukgi.

Figure E-2 Taegeukgi with four earthly trigrams removed

If we move the design of Figure E-2 into a 3 by 2 rectangle, the current Taegeukgi is obtained as seen in Figure E-3. If you understand what I have said consistently throughout this book, you will agree with the argument that the current Taegeukgi with four heavenly trigrams should be replaced by the original one with Bokhui's eight trigrams that is illustrated in Figure E-1 and was first proposed by Emperor Gojong. The perfect symbol of Eastern cosmology would then become our proud national flag. I personally think it would be wonderful for North Korea to play a role in adding the four earthly trigrams when we return to the original Taegeukgi of Figure E-1 to celebrate the reunification of the two Koreas.

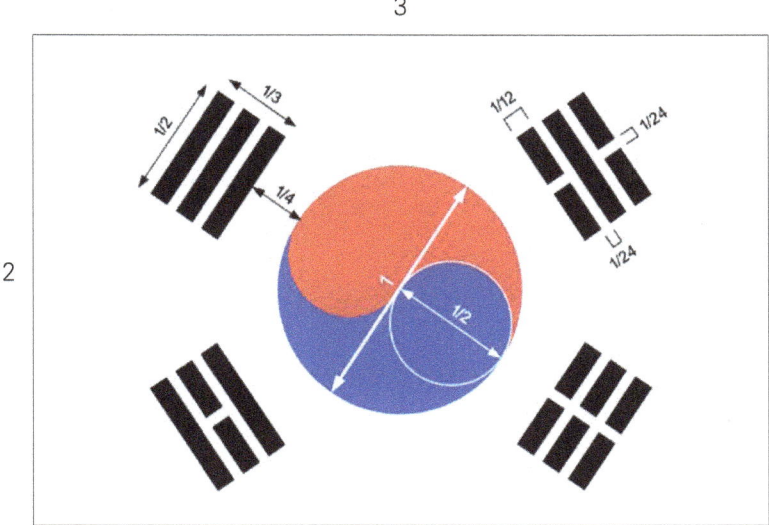

Figure E-3 Current Taegeukgi with four heavenly trigrams and how to draw it

However, such a fantastic national flag would be suitable only for those who understand at least what I am saying in this book. Of course, many people could complain that it is difficult to draw the new flag, Figure E-4. Some may object that the current flag is already difficult to draw, and the new one even more so. However, instead of such a blanket opposition, those people need to study the history of our national flag and its principles. Only good people deserve a good national flag. There are many national flags that are impossible to manually draw. If the national flag becomes easy to draw when one understands simple principles, then we need to teach those principles in elementary schools.

I would like to remind you that Taeho Bokhui, the originator of Taegeukgi, lived 5,500 years ago. Thus, our national flag is as much as 5,500 years old.

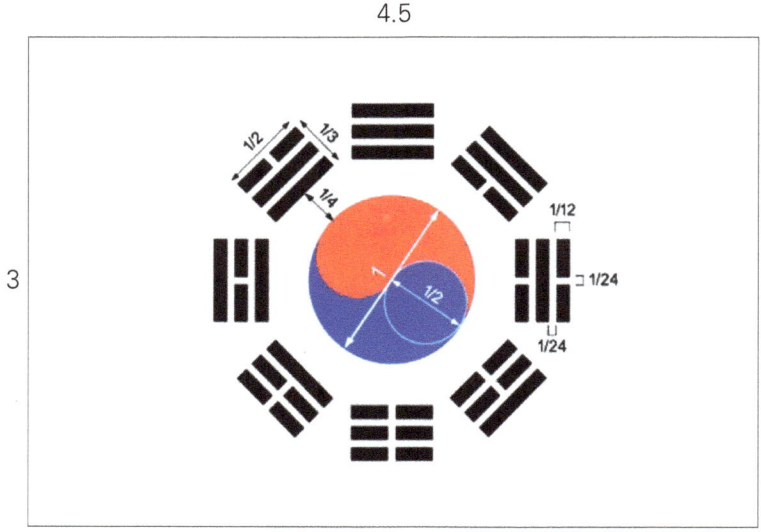

Figure E-4 Taegeukgi desirable for unified Korea

Finally, I heartily thank Jisook Chai, Ui Cheol Choi, Hyunjin Han, Gyeong Up Kang, Jae Nam Kim, and Taeho Kim for their devotion to publishing this book.

www.ingramcontent.com/pod-product-compliance
Lightning Source LLC
Chambersburg PA
CBHW042128160426
43198CB00021B/2938